AREA 51

First published in 2014 by Zenith Press, an imprint of Quarto Publishing Group USA Inc., 400 First Avenue North, Suite 400, Minneapolis, MN 55401 USA

ISBN: 978-0-7603-4664-8

Edited by Howard Zimmerman
Written by Dwight Zimmerman
Illustrated by Greg Scott
Cover and interior designed by Dean Motter

THIS IS A Z FILE INC. BOOK.

Printed in China

10 9 8 7 6 5 4 3 2 1

AREA 51

THE GRAPHIC HISTORY OF AMERICA'S MOST SECRET MILITARY INSTALLATION

DWIGHT JON ZIMMERMAN
NEW YORK TIMES BEST-SELLING AUTHOR
ILLUSTRATED BY GREG SCOTT

ZENITH PRESS

Contents

DAWN, EARLY 1980s. SOMEWHERE IN GROOM RANGE, NEVADA.
WHATEVER THIS FRIGGIN' THING IS-- I GOT IT!
THE GOVERNMENT IS SECRETLY IN LEAGUE WITH ALIENS! WE'VE GOT TO PUBLISH THIS--
THAT'S A SPACESHIP! THIS IS PROOF!
BUT WE NEED TO CLEAR OUT BEFORE ANY GUARDS FIND US!
SORRY, BOYS. THIS LAND IS POSTED. YOU HAVE TO CLEAR OUT ... NOW!

I SUGGEST YOU HEAD BACK TO YOUR VEHICLE.
YOU CAN'T KEEP WHAT'S GOING ON IN AREA 51 A SECRET FOREVER!
MISTER, I HAVE NO IDEA WHAT YOU'RE TALKING ABOUT.

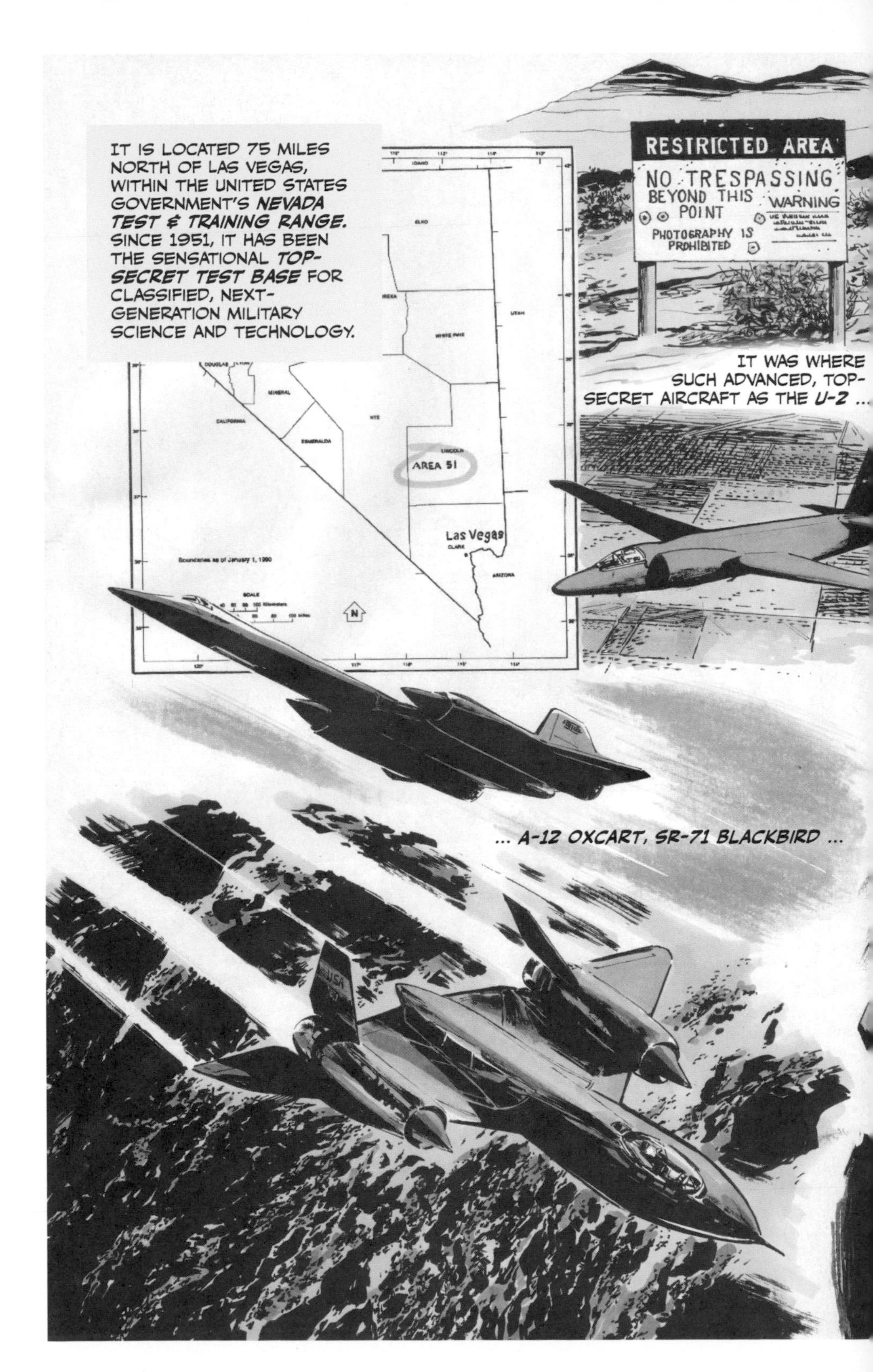
IT IS LOCATED 75 MILES NORTH OF LAS VEGAS, WITHIN THE UNITED STATES GOVERNMENT'S *NEVADA TEST & TRAINING RANGE.* SINCE 1951, IT HAS BEEN THE SENSATIONAL *TOP-SECRET TEST BASE* FOR CLASSIFIED, NEXT-GENERATION MILITARY SCIENCE AND TECHNOLOGY.
AREA 51
Las Vegas
RESTRICTED AREA
NO TRESPASSING BEYOND THIS POINT
WARNING
PHOTOGRAPHY IS PROHIBITED
IT WAS WHERE SUCH ADVANCED, TOP-SECRET AIRCRAFT AS THE *U-2* ...
... *A-12 OXCART, SR-71 BLACKBIRD* ...

... AND THE WORLD'S FIRST STEALTH FIGHTER JET, THE F-117 NIGHTHAWK, WERE ALL TESTED.
BUT, ECAUSE THE GOVERNMENT CONSISTENTLY *DENIED* AREA 51'S EXISTENCE, IT BECAME THE LOCUS FOR *CONSPIRACY THEORIES*--THE BIG THREE BEING ...
THE *LUNAR LANDING HOAX*
UFOS, ALIEN BODIES, AND CAPTURED *LIVE ALIENS,* HIDDEN IN A SECRET UNDERGROUND COMPLEX
AND A GIGANTIC *UNDERGROUND BUNKER AND TUNNEL SYSTEM,* CONNECTING MILITARY FACILITIES AND NUCLEAR LABORATORIES FROM *EAST COAST* TO *WEST COAST.*

A LOCAL TOLD ME ABOUT A TRAIL WE CAN USE TO GET TO A DIFFERENT RIDGE FOR OBSERVATION. I'LL GET THE SPARE CAMERA AND--
UH, GUYS? I DON'T LIKE THE LOOK OF THAT.
AWWWW, SH--
LOOK OUT!
K-POW!
I THINK WE'D BETTER JUST CLEAR OUT BEFORE SOMETHING ELSE HAPPENS.
IN JULY 2013, THE CIA FINALLY DECLASSIFIED AREA 51'S EXISTENCE AND RELEASED THOUSANDS OF DOCUMENTS RELATED TO ITS HISTORY.
AFTER 62 YEARS OF DENIAL, THE TRUE STORY OF THE GOVERNMENT'S MOST FAMOUS, AND MOST NOTORIOUS, SECRET FACILITY CAN FINALLY BE TOLD.

CHAPTER ONE
THE ORIGIN OF AREA 51: FROM ROSWELL TO OXCART
KIRTLAND AIR FORCE BASE, NEW MEXICO, NORTH OF THE WHITE SANDS PROVING GROUND, EARLY JUNE 1947...
WHAT THE HELL? CAPTAIN-- YOU'VE GOTTA SEE THIS!
WHAT DO YOU HAVE, SERGEANT?
TWO HIGH-PERFORMANCE AIRCRAFT JUST ENTERED OUR AIRSPACE--
--CLOCKED THEM AT AIR SPEEDS UP TO 1,200 MILES PER HOUR... WITH HOVER CAPABILITY!
IDENTITY?
UNKNOWN! DEFINITELY NOT OURS!
I NEED TO ALERT THE C.O.!*
THEY'RE... GONE!
*COMMANDING OFFICER

CAPTAIN KENNETH CHANDLER WAS ORDERED TO TAKE TO THE AIR AND INTERCEPT THE UNIDENTIFIED INTRUDER.
HE DIDN'T FIND ANYTHING.

BUT, A FEW DAYS LATER, WILLIAM "MACK" BRAZEL, A FOREMAN WORKING ON THE FOSTER HOMESTEAD ABOUT 30 MILES NORTH OF ROSWELL, NEW MEXICO, DID.
WHAT KINDA AIRCRAFT CRASHED HERE?

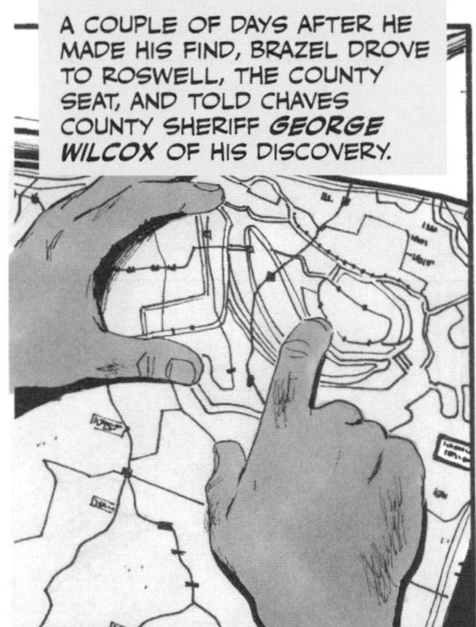
A COUPLE OF DAYS AFTER HE MADE HIS FIND, BRAZEL DROVE TO ROSWELL, THE COUNTY SEAT, AND TOLD CHAVES COUNTY SHERIFF ***GEORGE WILCOX*** OF HIS DISCOVERY.

THERE'S A LOT OF WRECKAGE, SHERIFF, SCATTERED OVER AT LEAST A HALF-MILE.
THANKS, MACK. I'LL LOOK INTO IT.

WILCOX CONTACTED ***MAJOR JESSE MARCEL,*** INTELLIGENCE OFFICER FOR THE 509TH BOMB GROUP STATIONED AT ROSWELL ARMY AIR FIELD, WHO WENT TO INVESTIGATE.

THE CRASH SITE WAS IMMEDIATELY SEALED UNTIL RETRIEVAL TEAMS COULD CLEAR THE AREA.
RESTRICTED AREA
NO UNAUTHORIZED
PERSONNEL ALLOWED

ON THE AFTERNOON OF JULY 8, 1947, *1ST LIEUTENANT WALTER HAUT,* 509TH OPERATIONS GROUP PUBLIC INFORMATION OFFICER AT ROSWELL ARMY AIR FIELD, HANDED KGFL RADIO ANNOUNCER FRANK JOYCE A PRESS RELEASE.
THIS CONTAINS EVERYTHING WE KNOW ABOUT THE *"FLYING DISK"* FOUND ON THE FOSTER HOMESTEAD.

THREE HOURS LATER, HAUT RETURNED WITH A SECOND PRESS RELEASE.

FORGET THAT FIRST ONE. IT HAD INCORRECT INFORMATION. USE THIS *CORRECTED* PRESS RELEASE.
TURNS OUT IT WAS WRECKAGE FROM A WEATHER BALLOON.
WILL DO, LIEUTENANT.
EIGHTH AIR FORCE COMMANDER *BRIGADIER GENERAL ROGER RAMEY,* UNDER WHOSE AUTHORITY THE ROSWELL AREA FELL, QUICKLY HELD A PRESS CONFERENCE, SHOWING MATERIAL HE CLAIMED WAS FOUND ON THE CRASH SITE.
AS YOU CAN SEE, IT'S MATERIAL FOR A SPECIAL WEATHER BALLOON.

AS THE AIR FORCE HOPED, THE STORY OF THE ROSWELL INCIDENT FADED AND, FOR THIRTY YEARS, WAS FORGOTTEN. BUT THE BUZZ AND HYSTERIA ABOUT UFOs ONLY GREW. TOO MANY PEOPLE FROM ALL WALKS OF LIFE AND IN TOO MANY PLACES HAD SEEN FLYING OBJECTS THAT DEFIED EASY EXPLANATION.
IN JULY 1947, WHEN THE UFO CRAZE (*AS IT CAME TO BE KNOWN*) BEGAN, 850 UFO SIGHTINGS WERE OFFICIALLY RECORDED.
OF THEM, MILITARY INTELLIGENCE FOUND 150 IMPORTANT ENOUGH TO INVESTIGATE.
TO DEAL WITH THE SITUATION, THE AIR FORCE CREATED TWO PROGRAMS: ***PROJECT SIGN,*** WHICH INVESTIGATED ACCOUNTS OF UFO SIGHTINGS ...
... AND ***PROJECT GRUDGE,*** A PUBLIC RELATIONS CAMPAIGN TO DEBUNK THE EXISTENCE OF UFOs, WHICH, AS FAR AS EVERYONE WAS CONCERNED, MEANT SPACECRAFT PILOTED BY BEINGS FROM ANOTHER PLANET.

AT WORLD WAR II'S END, TWO SUPERPOWERS EMERGED: THE UNITED STATES, WHICH HAD SUFFERED LITTLE DURING THE CONFLICT, AND THE SOVIET UNION, WHICH HAD BEEN SAVAGED BY IT.
THEIR VASTLY DIFFERENT POLITICAL SYSTEMS--DEMOCRACY AND COMMUNISM--MADE IT INEVITABLE THAT THE FORMER WORLD WAR II ALLIES WOULD BECOME ENEMIES.
THAT RUPTURE OCCURRED IN 1947 WITH THE BEGINNING OF THE COLD WAR, A WORLDWIDE IDEOLOGICAL STRUGGLE THAT MANY FEARED WOULD ERUPT INTO WORLD WAR III.
THIS WAS ALSO THE DAWN OF THE ATOMIC AGE.
TO DETER THE RUSSIANS, AMERICA'S MILITARY WAS CONDUCTING ABOVE-GROUND TESTS OF ATOMIC BOMBS, LIKE *OPERATION CROSSROADS* AT BIKINI ATOLL.
THESE BOMBS WERE THOUSANDS OF TIMES MORE POWERFUL THAN THOSE DROPPED ON HIROSHIMA AND NAGASAKI.

ЛЕВОЕ
СОПЛО
IN HER BOOK *AREA 51,*
ANNIE JACOBSEN ASSERTS
THE ROSWELL INCIDENT
WRECKAGE WAS A FLYING
DISK BUILT BY THE ***RUSSIANS!***

THE U.S. GOVERNMENT
HAS RELEASED
AT LEAST TWO
"OFFICIAL"
EXPLANATIONS OF
THE ROSWELL
CRASH. NEITHER
ONE SUPPORTS
JOHNSON'S
SCENARIO.
ЛЕВОЕ СОПЛО
ВЕРХ

THE TRUTH MAY
NEVER BE KNOWN.

ON AUGUST 29, 1949, THE SOVIET UNION
ANNOUNCED ITS FIRST SUCCESSFUL ATOMIC
BOMB TEST.

THE RUSSIANS
CREATED AN
"IRON CURTAIN"
OF PUPPET
STATES
CONTAINING
RED ARMY AND
RED AIR FORCE
BASES. THE
NEED FOR
INTELLIGENCE
ON SOVIET
MILITARY
CAPABILITY
BECAME
URGENT.
U.S.S.R.
GDR
Poland
Czechoslovakia
Hungary
Romania
Bulgaria

THE CLOSED POLICE STATE OF
THE SOVIET UNION MADE *HUMINT*--
HUMAN INTELLIGENCE, OR SPIES--
IMPRACTICAL AND DANGEROUS.
Щ 282

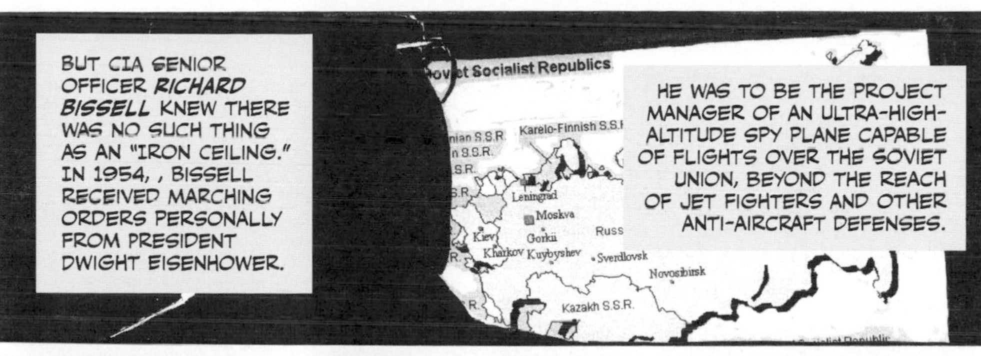
BUT CIA SENIOR OFFICER *RICHARD BISSELL* KNEW THERE WAS NO SUCH THING AS AN "IRON CEILING." IN 1954, , BISSELL RECEIVED MARCHING ORDERS PERSONALLY FROM PRESIDENT DWIGHT EISENHOWER.
HE WAS TO BE THE PROJECT MANAGER OF AN ULTRA-HIGH-ALTITUDE SPY PLANE CAPABLE OF FLIGHTS OVER THE SOVIET UNION, BEYOND THE REACH OF JET FIGHTERS AND OTHER ANTI-AIRCRAFT DEFENSES.
Leningrad
Moskva
Gorkii
Kuybyshev
Kiev
Kharkov
Sverdlovsk
Novosibirsk
Kazakh S.S.R.
Karelo-Finnish S.S.R.

THE AERODYNAMIC CHALLENGES TO CREATING SUCH AN AIRCRAFT WERE EXTRAORDINARY.

THE PHOTOS FROM MISSIONS OVERFLYING RUSSIA WOULD BE INVALUABLE!
BUT BISSELL KNEW THE MAN WHO COULD DO IT. *CLARENCE "KELLY" JOHNSON* WAS AMERICA'S TOP AIRPLANE DESIGNER.

KELLY? I HAVE A LITTLE PROJECT IN MIND FOR YOU.
YES, TOP SECRET.

KELLY'S SPECIAL FACILITY AT LOCKHEED, NICKNAMED *"THE SKUNK WORKS,"* WAS ALREADY FAMOUS FOR ITS MILITARY AIRCRAFT DESIGNS.
IF ANYONE COULD BUILD THE SPY PLANE--WHICH WOULD BEAR THE CODE NAME U-2--KELLY AND HIS TEAM COULD.

IN THE WINTER OF 1955, WITH DESIGN WORK ON THE U-2 NEARING COMPLETION, IT WAS TIME TO FIND A SECURE TEST SITE. BISSELL AND FELLOW CIA OFFICER HERBERT MILLER MET KELLY JOHNSON AND LOCKHEED TEST PILOT TONY LEVIER FOR A SPECIAL TRIP.
HI, RICK, HERB.
HELLO, KELLY.
EVERYONE READY FOR A LITTLE PLANE RIDE TO NOWHERE?
THE TEST SITE'S REQUIREMENT POSED A CHALLENGE. IT HAD TO BE REMOTE, YET ALSO CLOSE TO THE SKUNK WORKS.
LEVIER HAD WHAT HE THOUGHT WAS A GOOD CANDIDATE, ONE HE AND FELLOW LOCKHEED TEST PILOT RAY GOUDY HAD VISITED IN A PRE-SCOUTING MISSION.
THERE IT IS, FELLAS.
GROOM LAKE, NEVADA.

YOU CAN SEE IT WAS LAID OUT WITH A TEMPORARY LANDING STRIP--FOR EMERGENCIES DURING PILOT TRAINING AT *NELLIS* IN WORLD WAR II.
IS THIS INSIDE THE *NEVADA TEST SITE* BOUNDARIES?
ACTUALLY, WE'RE JUST EAST OF WHERE THEY DO THE ATOMIC BOMB TESTS.
THIS PLACE IS PERFECT. I'M RECOMMENDING THE PRESIDENT ADD IT TO THE NEVADA TEST SITE.
THAT WAY WE'LL ALSO HAVE COVER, AND SECURITY, FROM THE ATOMIC ENERGY COMMISSION.
AREA 51'S FALLOW PERIOD WAS OVER. FOUR MONTHS AFTER THEIR FLYOVER, GROOM LAKE WAS IN BUSINESS.
DOUGLAS C-124 GLOBEMASTERS CARRIED DISASSEMBLED U-2s FROM BURBANK TO GROOM LAKE.
THE PIECES, ALL WRAPPED IN SHEETS TO FURTHER DISGUISE THEM, WERE CARRIED INTO THE HANGER WHERE THE U-2s WERE ASSEMBLED.

ONLY WHEN IT WAS READY TO FLY WOULD THE U-2 EMERGE.
THE U-2 WAS A UNIQUE AIRCRAFT, HAVING A SERVICE CEILING OF 70,000 FEET AND A RANGE OF 5,500 NAUTICAL MILES AND BEING CAPABLE OF SUSTAINED FLIGHT FOR NINE HOURS.
TEST PILOT *RAY GOUDY* WAS THE FIRST TO REACH 65,000 FEET. THE VIEW WAS SPECTACULAR. HE WAS ABLE TO SEE THE PACIFIC OCEAN, 300 MILES AWAY.
BUT BECAUSE THE U-2 HAD NOT YET BEEN PAINTED BLACK, ITS ALUMINUM FUSELAGE AND WINGS ACTED LIKE MIRRORS REFLECTING SUNLIGHT...
... MAKING IT VISIBLE TO AIRCRAFT FLYING 40,000 FEET LOWER.
TRANS WORLD AIRLINES
TWA
CAPTAIN! AT OUR TWO O'CLOCK HIGH --VERY HIGH.
WHAT DO YOU SUPPOSE THAT IS?
NO IDEA. BUT IT'S NOT A METEORITE-- ITS FLIGHT IS TOO CONTROLLED.

AS THE U-2 FLIGHTS CONTINUED, UFO SIGHTINGS IN CALIFORNIA, NEVADA, AND UTAH SPIKED. COMMERCIAL PILOTS AND AIR TRAFFIC CONTROLLERS FLOODED CIA HEADQUARTERS WITH PHONE CALLS.
MORE THAN HALF THE UFO SIGHTINGS INVOLVED U-2 FLIGHTS. BUT THE CIA AND THE AIR FORCE REFUSED TO REVEAL THE EXISTENCE OF THE U-2 PROGRAM BECAUSE THEY DIDN'T WANT THE RUSSIANS TO KNOW ABOUT IT.

GOVERNMENT STONEWALLING AND OBFUSCATION WITH *PROJECT SIGN*--ITS OFFICIALLY ANNOUNCED LOOK INTO UFO PHENOMENA--AND *PROJECT GRUDGE,* A SIMILAR PROGRAM STARTED BY THE AIR FORCE, SIMPLY ADDED FUEL TO THE UFO CONSPIRACY FIRE.
CIA CHIEF ***ALLEN DULLES*** IGNORED QUERIES FROM PRIVATE CITIZENS. ANY UFO GROUP CONTACTING THE CIA WAS MONITORED. AND SENATORS AND CONGRESSMEN WHO INQUIRED RECEIVED A LETTER STATING UFOs WERE A LAW ENFORCEMENT PROBLEM, SOMETHING BEYOND THE LEGAL AUTHORITY OF THE CIA.
UFO'S ARE REAL!
TELL THE TRUTH ABOUT UFO'S

ON JULY 4, 1956, FROM A SECRET U-2 BASE NEAR WIESBADEN, WEST GERMANY, HERVEY STOCKMAN LIFTED HIS U-2 INTO THE SKY...
... AND BECAME THE FIRST MAN TO PENETRATE SOVIET AIRSPACE, PHOTOGRAPHING MILITARY FACILITIES IN AND AROUND LENINGRAD (NOW ST. PETERSBURG).
THE PHOTOS COVERED MORE THAN 400,000 SQUARE MILES AND PRODUCED A GOLD MINE OF SOLID INFORMATION.
FOR THE FIRST TIME, WE ACTUALLY KNOW WHAT'S GOING ON WITHIN THE BORDERS OF THE SOVIET UNION.
BUT THE U-2 WAS NOT A STEALTH AIRCRAFT. SOVIET RADAR WAS ABLE TO TRACK STOCKMAN'S AND ALL OTHER U-2 FLIGHTS. TO THE SOVIET LEADERS' FURY, HOWEVER THE U-2 WAS BEYOND THE RANGE OFTHEIR DEFENSES ... FOR NOW.

TO COUNTER SOVIET RADAR, BISSELL ORDERED U-2s TO BE COATED WITH RADAR-ABSORBING PAINT.
THOSE TESTS ENDED IN FAILURE, WITH TEST PILOT *ROBERT SIEKER* DYING IN A CRASH NEAR GROOM LAKE IN APRIL 1957.

AND THE U-2'S INTELLIGENCE SUCCESS WAS GIVING *PRESIDENT DWIGHT EISENHOWER* DIPLOMATIC HEADACHES.
MAYBE IT'S TIME TO SHUT DOWN AREA 51.
BISSELL KNEW AREA 51 WAS IN TROUBLE. BUT, BRAZENLY, HE DECIDED TO DOUBLE DOWN WITH A SUCCESSOR TO THE U-2, A PLANE EVEN MORE TECHNOLOGICALLY ADVANCED, CAPABLE OF FLYING AT 90,000 FEET AT A SPEED OF MACH 3 AND WITH STEALTH CAPABILITY.
IT WOULD BE A BILLION-DOLLAR GAMBLE. WOULD PRESIDENT EISENHOWER APPROVE IT?

ON OCTOBER 4, 1957, THE SOVIET UNION ONCE AGAIN STUNNED THE WORLD--THIS TIME, BY SENDING INTO ORBIT *SPUTNIK 1*, A MAN-MADE SATELLITE THAT CIRCLED THE EARTH FOR 21 DAYS.

BY BEING THE FIRST TO SEND A MAN-MADE SATELLITE INTO SPACE, SOVIET ROCKET SCIENTISTS, WORKING SECRETLY, HAD TRUMPED AND EMBARRASSED THE PUBLIC AMERICAN ROCKET PROGRAM, WHICH HAD SUFFERED ONE SPECTACULAR FAILURE AFTER ANOTHER.
SPUTNIK CREATED A CRISIS OF CONFIDENCE IN THE COUNTRY. EISENHOWER MADE ***JAMES KILLIAN*** SPECIAL ASSISTANT TO THE PRESIDENT FOR SCIENCE AND TECHNOLOGY.

KILLIAN AND BISSELL WERE LONG-STANDING FRIENDS.
JIM, CONGRATULATIONS ON YOUR NEW JOB. NOW, LET ME TELL YOU ABOUT THIS PROJECT I HAVE IN MIND.
IF IT WILL PUT US AHEAD OF THE RUSSIANS, I'M ALL FOR IT.
THAT IT WILL.

BISSELL LAID OUT HIS PLANS FOR THE ULTIMATE SPY PLANE, CODENAMED ***AQUATONE,*** THAT WOULD BE GENERATIONS AHEAD OF ANYTHING THE RUSSIANS COULD PRODUCE.
KILLIAN WAS IMPRESSED AND GOT PRESIDENT EISENHOWER TO APPROVE IT. AREA 51 HAD A STAY OF EXECUTION. FOR IT TO GET BACK INTO BUSINESS, IT NEEDED AQUATONE TO WORK.

KELLY JOHNSON WANTED THE AQUATONE CONTRACT. HE NEEDED SOMEONE WHO COULD STUMP RADAR, AND FOR THAT, IN SEPTEMBER 1957, HE WENT TO LOCKHEED'S RADAR MAN, EDWARD LOVICK.
ED, WOULD YOU LIKE TO COME WORK ON AN INTERESTING PROJECT?
IF IT MEANS WORKING AGAIN WITH MacDONALD AND MY OLD BOSS BILL MARTIN, SURE.
GREAT. ONCE YOU GET YOUR TOP-SECRET SECURITY CLEARANCE, WE'LL TELL YOU MORE.
LOVICK SOON FOUND HIMSELF AT AREA 51, TRAILBLAZING A TECHNOLOGY THAT WOULD MAKE HIM KNOWN AS THE "GRANDFATHER OF STEALTH." HE SUCCESSFULLY REDUCED THE A-12'S RADAR SIGNATURE TO THAT OF A SMALL BIRD.
JOHNSON GOT THE CONTRACT, AND THE PROJECT GOT A NEW CODENAME: OXCART.
ON MAY 1, 1960, THE SOVIET UNION FINALLY HAD SUCCESS AGAINST THE U-2 WHEN IT SHOT DOWN ONE, PILOTED BY FRANCIS GARY POWERS FLYING AT 70,000 FEET. A NEW CHAPTER IN THE HISTORY OF AREA 51 WAS ABOUT TO BEGIN.
POWERS AND THE WRECKAGE OF HIS SPY PLANE WERE RETRIEVED BY THE SOVIETS, WHO PARADED BOTH BEFORE THE ENTIRE WORLD. IT WAS A LOW MOMENT FOR AMERICAN COLD WARRIORS.

THE NATIONAL SECURITY PROBLEM CONFRONTING THE CIA AND THE PENTAGON DURING THE COLD WAR WAS SIMPLE. THEY NEEDED TO KNOW WHAT THEIR ADVERSARIES--THE RUSSIANS, THE CHINESE, THE NORTH KOREANS--WERE UP TO. THE SITUATION WAS COMPLEX.

AT OPPOSITE ENDS WERE MANNED AIRCRAFT, WHICH COULD RESPOND TO REAL-TIME NEEDS BUT WITH DIRE POLITICAL CONSEQUENCES SHOULD ANOTHER SPY PLANE BE SUCCESSFULLY SHOT DOWN ...
... AND SATELLITES, WHOSE SHOOT-DOWN RISK WAS ZERO BUT WHOSE FLIGHT PATHS COULD BE OBSERVED AND TIMED.
SOMEWHERE IN THE MIDDLE WERE THE RPVs AND UAVs.
THE TESTING OF THE MANY MANNED AND UNMANNED AIRCRAFT SYSTEMS IN THE 1960S AND 1970S WOULD TRANSFORM AREA 51 INTO A BUSTLING, COMPARTMENTALIZED, TOP-SECRET GRAND CENTRAL TERMINAL.
THE PARTICIPANTS WERE AWARE OF THE HEIGHTENED ACTIVITY, BUT UNLESS ON NEED-TO-KNOW SHORT LISTS, THEY WERE IGNORANT OF ITS NATURE.

CHAPTER TWO

FROM OXCART TO THE DAWN OF THE DRONES

WORK CREWS FROM THE ADJACENT NEVADA TEST SITE, ALREADY POSSESSING NECESSARY TOP-SECRET CLEARANCES, BEGAN CONSTRUCTION IN SEPTEMBER 1960. FOR OXCART, THE BASE'S ENTIRE INFRASTRUCTURE HAD TO BE EXPANDED AND, IN SOME CASES, REDESIGNED.

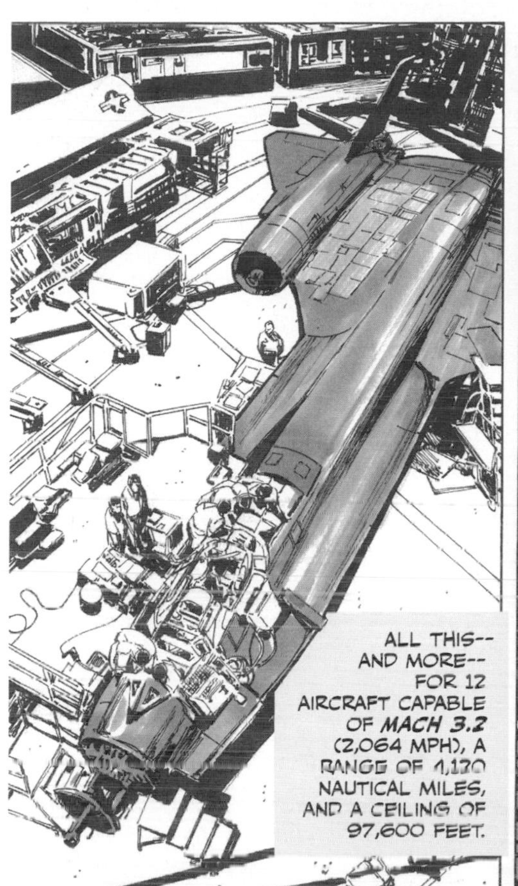

A DECISION WAS MADE TO USE THE SMALLER P&W J75 UNTIL THE BUGS IN THE J58 WERE SOLVED.

DELIVERY WAS PUSHED BACK TO AUGUST. THEN AGAIN, TO DECEMBER.

ON THE MORNING OF APRIL 25, 1962 ...
LOU, WAKE UP!

TIME TO FLY.

AS LOCKHEED TEST PILOT ***LOU SCHALK*** CLIMBED INTO THE COCKPIT OF THE A-12.

THE SMALL GROUP OF OBSERVERS IN THE CONTROL TOWER HELD THEIR BREATH. MILLIONS OF DOLLARS HAD BEEN SPENT TO CREATE AN AIRPLANE TECHNOLOGICALLY 40 YEARS AHEAD OF ITS TIME.
NOW, WOULD IT WORK?

SCHALK WASN'T SCHEDULED TO FLY THE A-12 THAT DAY, JUST CONDUCT A TAXI TEXT.

... RICHARD BISSELL, THE TRIUMPH WAS BITTER-SWEET. HE WAS HERE NOT AS AREA 51'S BOSS, BUT AS KELLY JOHNSON'S GUEST.

BISSELL HAD BEEN ONE OF THE KEY PLANNERS OF THE MILITARY DISASTER KNOWN AS THE *BAY OF PIGS* INVASION--A CIA-BACKED EFFORT USING CUBAN EXILES TO ATTEMPT TO OVERTHROW FIDEL CASTRO'S COMMUNIST GOVERNMENT.
THE INVASION WAS A SPECTACULAR FAILURE.
INSTEAD OF ACCEPTING A DEMOTION, BISSELL RESIGNED. JOHNSON'S INVITATION WAS A GESTURE OF LOYALTY TO A FRIEND. IT WOULD BE THE LAST TIME BISSELL SET FOOT ON THE FACILITY HE CREATED.
ALL THE CONSTRUCTION AND ACTIVITY AT AREA 51 WAS CLOSELY MONITORED--NOT BY AMERICAN CITIZENS, BUT BY THE RUSSIANS, WHOSE INTEREST IN AMERICAN SPY-PLANE DEVELOPMENT WAS MANIFEST.
FIRST WITH THE *OBJECT D* (ALSO KNOWN AS SPUTNIK 3), AND LATER WITH ITS *ZENIT* SATELLITES, THE RUSSIANS WERE MAKING REGULAR PASSES OVER AREA 51.
BECAUSE THE ORBITAL OVERFLIGHTS OPERATED ON A FIXED SCHEDULE, A-12 TESTS WERE CONDUCTED DURING THOSE PERIODS WHEN THE SATELLITES WERE KNOWN NOT TO BE OVERHEAD. THEN, ONE DAY...
INTELLIGENCE
IT'S A SKETCH OF THE A-12! DID THE RUSSIANS PLANT A SPY?
THE CIA LATER DETERMINED THAT SOME RUSSIAN SATELLITES HAD INFRARED CAPABILITY AND HAD PHOTOGRAPHED THE HEAT SIGNATURE OF AN A-12 LEFT ON THE DESERT FLOOR.

KEEPING THE AMERICAN PUBLIC IGNORANT OF AREA 51 AND ITS PURPOSE GENERATED INTERNAL DEBATES REGARDING THE OFTEN CONFLICTING NEEDS OF NATIONAL SECURITY AND PUBLIC TRUST IN THE PRESIDENT.

THE U-2 SHOOTDOWN AND THE BAY OF PIGS INCIDENTS HAD DAMAGED THE PUBLIC'S FAITH IN TWO PRESIDENTS, EISENHOWER AND KENNEDY, AND THE CIA.

IF THE PUBLIC WERE TO LEARN A MACH 3 SPY-PLANE PROGRAM DESIGNED TO PENETRATE RUSSIAN AIRSPACE IS ONGOING DESPITE PRESIDENTIAL ASSURANCES OTHERWISE ...
...THERE'LL BE WORSE ***HELL TO PAY*** THAN EITHER THE U-2 SHOOTDOWN OR THE BAY OF PIGS FIASCO.
WE MUST CONTINUE TO DENY AREA 51'S EXISTENCE--EVEN ITS NAME CANNOT BE MENTIONED.
WHAT ABOUT UFO SIGHTINGS? THEY'RE STARTING TO COME IN.
DENY AND STONEWALL.

IN THE POWER VACUUM FOLLOWING BISSELL'S OUSTER, THE AIR FORCE BEGAN TAKING MORE AND MORE CONTROL OF AREA 51.

THANKS TO THE DIPLOMATIC EFFORTS OF ***BRIGADIER GENERAL JACK LEDFORD*** AND THE CIA'S ***ALBERT "BUD" WHEELON***, A TEMPORARY POWER-SHARING ARRANGEMENT WAS REACHED IN WHICH THE AIR FORCE RAN OPERATIONS AT AREA 51 AND THE CIA RAN THE MISSIONS.

IT HAD ITS OWN TOP-SECRET COUNTERPART, ***NII-88***, NORTH OF MOSCOW, WHERE SERGEI KOROLOV WAS THE CHIEF DESIGNER OF THEIR SPACE PROGRAM.
THE SOVIET UNION COULD HAVE BLOWN AREA 51'S COVER, BUT THE SECRETIVE, TOTALITARIAN GOVERNMENT HAD ITS OWN REASONS FOR ASSISTING IN AREA 51'S DENIABILITY.
COMRADE KHRUSCHEV*, REVELATION OF AREA 51 WOULD EXPOSE OUR OWN DEFICIENCY IN SPY-PLANE TECHNOLOGY.
MARSHAL VARENTSOV**, THIS IS SOMETHING WE CANNOT MAKE PUBLIC.
UNLESS THERE IS A BREACH WE CAN EXPLOIT, LIKE THE U-2 SHOOTDOWN, WE WILL KEEP KNOWLEDGE OF AREA 51 WITHIN THE "INTELLIGENCE FAMILY."
**NIKITA KHRUSCHEV*, PREMIER OF THE SOVIET UNION 1953–1964.
**CHIEF MARSHAL OF ARTILLERY SERGEY VARENTSOV, MISSILE EXPERT.

FROM 1962 TO 1968 A TOTAL OF 2,850 OXCART FLIGHTS FROM AREA 51 WERE CONDUCTED, CREATING A MAJOR UFO SIGHTING HEADACHE FOR THE CIA.

THE MOST SPECTACULAR AND WORRISOME OXCART-RELATED UFO SIGHTING WAS THE FIRST, OCCURRING ON APRIL 30, 1962, JUST FOUR DAYS AFTER OXCART'S FIRST OFFICIAL FLIGHT.
IT WAS PROBLEMATIC BECAUSE IT RESULTED FROM ANOTHER GOVERNMENT PROGRAM: NASA WAS IN THE MIDDLE OF TESTING THE X-15 ROCKET PLANE.

U.S. AIR FORCE
U.S. AIR FORCE
PART OF AMERICA'S SPACE PROGRAM, THE X-15 WAS LAUNCHED FROM BENEATH THE WING OF A MODIFIED B-52. IT WAS THE FIRST MANNED VEHICLE TO REACH THE EDGE OF SPACE.
PART OF TEST PILOT JOE WALKER'S MISSION WAS TO TAKE PHOTOGRAPHS OF EARTH.

AS THE X-15 WAS A NON-CLASSIFIED PROGRAM, WALKER'S PHOTOS WERE RELEASED TO THE PUBLIC.

IN THE CORNER OF ONE WAS A TINY IMAGE OF AN A-12.
CIA DENIALS ONLY FUELED MOUNTING UFO CONSPIRACY THEORIES.

ADDITIONAL MISSIONS REVEALED MORE SITES UNDER CONSTRUCTION, SPARKING INTENSE DEBATE BETWEEN THE PENTAGON AND THE CIA OVER BOTH OVERALL PURPOSE AND THE EXTENT OF PROGRESS.

THE KENNEDY ADMINISTRATION WAS ABLE TO SHOW THE WHOLE WORLD THE CUBAN MISSILE BASE, USING PHOTOS TAKEN FROM SPY PLANE OVERFLIGHTS. EVENTUALLY A DEAL WAS REACHED AND THE SOVIETS WITHDREW THEIR MISSILES.

GETTING OXCART OPERATIONAL BECAME A TOP PRIORITY.
BUT DELAYS WERE PILING UP. THERE WERE CRASHES, CAUSED BY EQUIPMENT FAILURE.

ELABORATE SECURITY MEASURES HAD TO BE USED WHEN SUCH CRASHES OCCURRED OUTSIDE AREA 51, ADDING TO THE DELAYS.

MEANWHILE, THE AIR FORCE DECIDED TO GET INTO THE ACT. IT ORDERED THREE A-12 VARIANTS, DESIGNATED RS-71.

THEY INCLUDED AN INTERCEPTOR CAPABLE OF CARRYING NUCLEAR BOMBS, A DRONE-CARRYING "MOTHER SHIP," AND A TWO-SEAT VERSION DESIGNED FOR NUCLEAR BOMB DAMAGE-ASSESSMENT PHOTORECONNAISSANCE.

I UNDERSTAND, SIR. THEN LET ME SUGGEST THIS: THE AIR FORCE HAS ITS VERSION OF THE A-12, THE *RS-71*.

SIMILAR PERFORMANCE, EXPANDED CAPABILITIES. THEY'RE STILL CONDUCTING TESTS, BUT WE CAN ROLL OUT ONE OF THEIRS, AND STILL KEEP OXCART SECRET.

AT A PRESS CONFERENCE ON FEBRUARY 29, 1964, PRESIDENT JOHNSON ANNOUNCED...
THE UNITED STATES HAS SUCCESSFULLY DEVELOPED AN ADVANCED EXPERIMENTAL JET AIRCRAFT, THE A-11, WHICH HAS BEEN TESTED IN SUSTAINED FLIGHT AT MORE THAN 2,000 MILES AN HOUR, AND AT ALTITUDES IN EXCESS OF 70,000 FEET.
THE DESIGNATION "A-11" WAS FALSE.
WHAT THE PRESS ACTUALLY SAW AT EDWARDS AIR FORCE BASE WAS A *YF-12*, THE EXPERIMENTAL DESIGNATION OF THE AIR FORCE INTERCEPTOR.
IT WAS ALL PART OF AN ELABORATE DECEPTION CREATED BY THE CIA: THE "OUTING" OF A SECRET WEAPON SYSTEM WHILE KEEPING THE TEST SITE ITSELF SECRET. ON APRIL 11, 1964, LBJ CONTINUED TO MIX TRUTH AND FALSEHOODS ABOUT OXCART.
THE WORLD RECORD FOR AIRCRAFT SPEED, CURRENTLY HELD BY THE SOVIETS, HAS BEEN REPEATEDLY BROKEN IN SECRECY BY THE UNITED STATES AIRCRAFT A-11...
THE SOVIET RECORD IS 1,665 MILES AN HOUR.
THE A-11 HAS ALREADY FLOWN IN EXCESS OF 2,000 MILES AN HOUR.
AND, ON JULY 24, 1964...
I WOULD LIKE TO ANNOUNCE THE SUCCESSFUL DEVELOPMENT OF A MAJOR NEW STRATEGIC AIRCRAFT...
THE *SR-71* AIRCRAFT...IS THE MOST ADVANCED IN THE WORLD. IT WILL FLY AT MORE THAN THREE TIMES THE SPEED OF SOUND...
WITHIN THE MILITARY, THE PRESIDENT, AS COMMANDER IN CHIEF, DOES NOT MAKE A MISTAKE. AS OF JULY 24, THE RS-71 WAS NOW THE *SR-71*.

BY 1966, OXCART WAS AT A CROSSROADS. WITH INCREASINGLY SOPHISTICATED SATELLITES LIKE CORONA COMING ONLINE, ITS ORIGINAL PURPOSE AS SUCCESSOR TO THE U-2 HAD ALL BUT ERODED.
SPY SATELLITES WERE RAPIDLY REPLACING SPY PLANES.

THE JOHNSON ADMINISTRATION FACED THE DUAL ESCALATING COSTS OF THE VIETNAM WAR AND HIS GREAT SOCIETY DOMESTIC PROGRAM. THE CIA'S TOP-SECRET OXCART PROGRAM HAD NO POLITICAL CLOUT AND WAS JUDGED AN EXPENDABLE LUXURY.

IN 1967, AFTER NINE YEARS AND MILLIONS OF DOLLARS SPENT, OXCART WAS SCHEDULED FOR RETIREMENT.

ON MAY 16, 1967, CIA DIRECTOR *RICHARD HELMS* MADE ONE LAST PITCH TO PRESIDENT JOHNSON TO USE OXCART BEFORE IT WAS MOTHBALLED.
WE CAN USE IT TO FIND THE MOBILE NORTH VIETNAMESE SURFACE-TO-AIR MISSILES THAT HAVE BEEN SAVAGING AMERICAN WARPLANES.
BUT IT HAS TO HAPPEN NOW--BEFORE THE MONSOON SEASON STARTS IN JUNE.
APPROVED. WHERE WILL YOU BASE THE SQUADRON?
THE 1129TH SPECIAL ACTIVITIES SQUADRON WILL OPERATE OUT OF KADENA, IN OKINAWA.

FIFTEEN DAYS LATER, CIA PILOT MELE VOJVODICH FLEW THE FIRST OXCART MISSION FOR *OPERATION BLACK SHIELD.*
PHOTOS FROM THE FLIGHT REVEALED 70 OF 190 SUSPECTED SAM SITES. A TOTAL OF 29 BLACK SHIELD MISSIONS WERE FLOWN. MOST WERE OVER NORTH VIETNAM, CAMBODIA, AND LAOS. TWO WERE IN THE OTHER DIRECTION.
CHINA
NORTH VIETNAM
HANOI
LAOS
Gulf of Tonkin
TRAVELING AT MACH 3.1 AT AN ALTITUDE OF 80,000 FEET, HIS FLIGHT OVER NORTH VIETNAM TOOK LESS THAN 10 MINUTES.
Phu Bai
Da Nang
THAILAND
Ubon
Chu Lai
Korat
SOUTH
Phu Cat
ON JANUARY 23, 1968, THE USS *PUEBLO*, AN ELECTRONIC INTELLIGENCE-GATHERING SHIP, WAS CAPTURED WHILE OPERATING OFF THE COAST OF NORTH KOREA.
THE JOHNSON ADMINISTRATION WAS CONCERNED NORTH KOREA WAS PREPARING TO LAUNCH A SECOND KOREAN WAR, SOMETHING AMERICA COULD NOT AFFORD.

BLACK SHIELD WOULD BE THE FIRST, AND LAST, MISSION FOR THE 1129TH SPECIAL ACTIVITIES SQUADRON, NICKNAMED THE "ROAD RUNNERS."

OF ALL THE BLACK--OR "SECRET"--PROGRAMS OPERATING AT AREA 51, ARGUABLY THE MOST SENSITIVE WAS *OPERATION HAVE DOUGHNUT*--AMERICA'S OWN SOVIET AIR FORCE.
THE OPERATION'S NAME, HAVE DOUGHNUT, WAS INSPIRED BY THE SHAPE OF THE MIG'S NOSE.
ITS FIRST PLANE, A TOP-OF-THE-LINE *MIG 21* FIGHTER FROM THE IRAQI AIR FORCE, ARRIVED IN 1967 COURTESY OF ISRAEL'S MOSSAD. THEY GOT IT WHEN AN IRAQI PILOT DEFECTED, ALONG WITH HIS PLANE, TO ISRAEL.
THE AGILE MIG-21s AND MIG-17s USED BY THE NORTH VIETNAMESE AIR FORCE, MANY FLOWN BY SOVIET PILOTS, WERE ACHIEVING 9:1 KILL RATIOS AGAINST THE MORE TECHNOLOGICALLY ADVANCED BUT LESS MANEUVERABLE AMERICAN WARPLANES.
WITH POSSESSION OF THE MIG-21, DESIGNERS COULD RE-ENGINEER THE AIRCRAFT AND PILOTS COULD GET INVALUABLE FIRST-HAND EXPERIENCE THROUGH FLYING THE AIRCRAFT AND DESIGN A DOCTRINE TO DEFEAT IT.

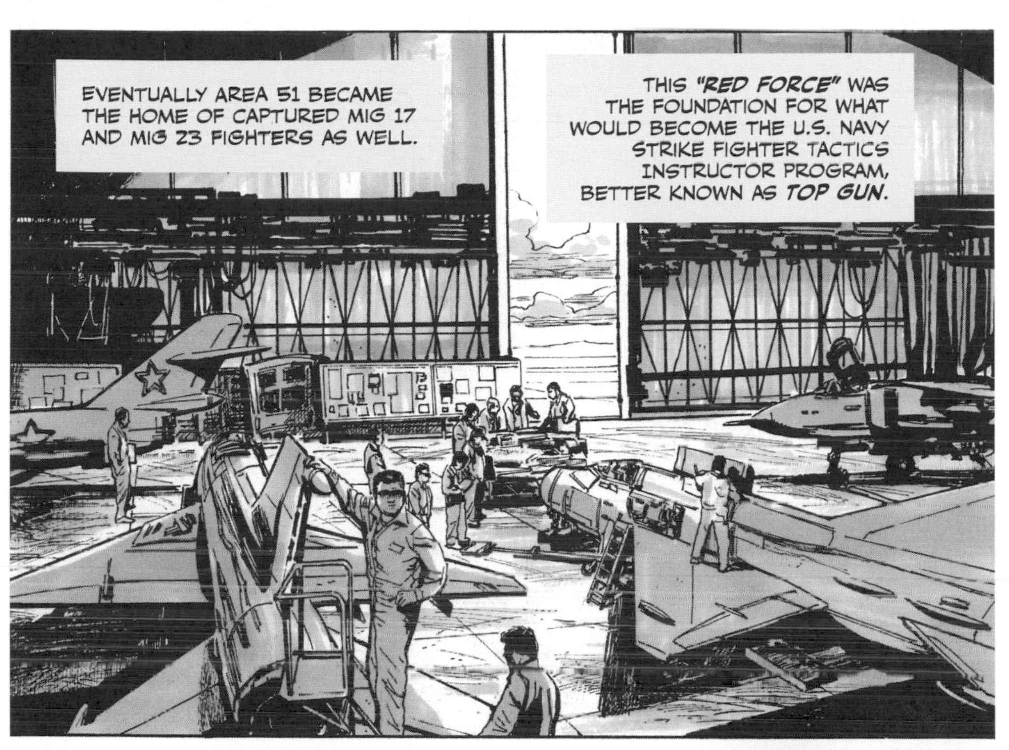
EVENTUALLY AREA 51 BECAME THE HOME OF CAPTURED MIG 17 AND MIG 23 FIGHTERS AS WELL.
THIS "RED FORCE" WAS THE FOUNDATION FOR WHAT WOULD BECOME THE U.S. NAVY STRIKE FIGHTER TACTICS INSTRUCTOR PROGRAM, BETTER KNOWN AS TOP GUN.

TALLY-HO. LAMBS FOR THE SLAUGHTER, 11 O'CLOCK LOW. RED THREE TAKE THE LEFT. RED TWO, THE RIGHT. I'VE GOT THE MIDDLE.
ROGER.
ROGER.

WELCOME TO TOP GUN, IMPERIALIST AMERICAN FIGHTER PILOTS! YOU HAVE ALL JUST BEEN SHOT DOWN!
WHAT THE HELL--?
WHERE'D YOU GUYS COME FROM--?

I BELIEVE IT WAS YOUR IMPERIALIST PRESIDENT ROOSEVELT WHO SAID IT BEST: **"SHANGRI-LA."**
IMPOSSIBLE!
SIREN SAYS OTHERWISE, IMPERIALIST DEAD DOG. AS DO OUR CAMERAS. WILL GREATLY ENJOY COLD BEER YOU BUY AT OFFICER CLUB.
THE DOGFIGHT WAS REAL; THE WEAPONS WERE NOT. WHEN A PLANE HAD AN ENEMY TARGETED IN ITS SIGHTS AND PULLED THE TRIGGER, A SCREECHING TONE WOULD SOUND IN THE VICTIM'S COCKPIT. IT MEANT THE TARGETED PLANE HAD JUST BEEN SHOT OUT OF THE AIR.

CHAPTER THREE

UNLEASH THE DRONES

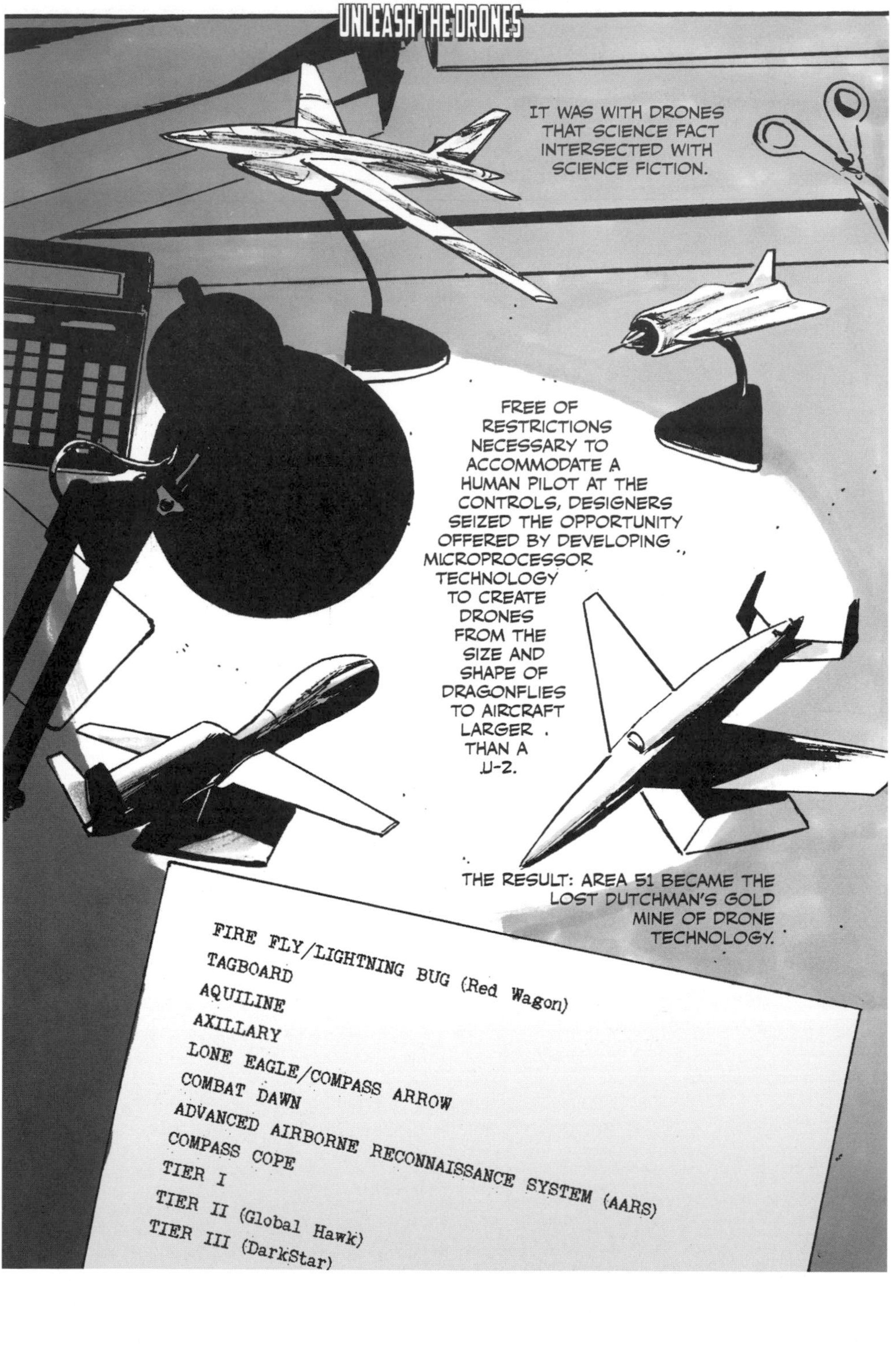

IT WAS WITH DRONES THAT SCIENCE FACT INTERSECTED WITH SCIENCE FICTION.

FREE OF RESTRICTIONS NECESSARY TO ACCOMMODATE A HUMAN PILOT AT THE CONTROLS, DESIGNERS SEIZED THE OPPORTUNITY OFFERED BY DEVELOPING MICROPROCESSOR TECHNOLOGY TO CREATE DRONES FROM THE SIZE AND SHAPE OF DRAGONFLIES TO AIRCRAFT LARGER THAN A U-2.

THE RESULT: AREA 51 BECAME THE LOST DUTCHMAN'S GOLD MINE OF DRONE TECHNOLOGY.

FIRE FLY/LIGHTNING BUG (Red Wagon)
TAGBOARD
AQUILINE
AXILLARY
LONE EAGLE/COMPASS ARROW
COMBAT DAWN
ADVANCED AIRBORNE RECONNAISSANCE SYSTEM (AARS)
COMPASS COPE
TIER I
TIER II (Global Hawk)
TIER III (DarkStar)

MILITARY INTELLIGENCE DRONE PROJECTS WERE FUNDED BY THE CLANDESTINE NATIONAL RECONNAISSANCE OFFICE. CREATED ON SEPTEMBER 6, 1961, EVEN ITS NAME WOULD REMAIN CLASSIFIED UNTIL 1992.
THE NRO (CODENAMED *BIG SAFARI*) DEVELOPED RED WAGON--RECONNAISSANCE DRONES BASED ON RYAN AERONAUTICAL'S "FIREBEE" TARGET DRONE.
IN ADDITION TO RECONNAISSANCE DRONES, BIG SAFARI DEVELOPED A RED WAGON VARIANT DESIGNED TO DETECT AND RELAY SAM SEARCH-AND-ACQUISITION RADAR SIGNALS.

SUCH MISSIONS, WHICH WOULD HAVE BEEN SUICIDAL FOR PILOTED AIRCRAFT, WERE INVALUABLE IN DEVELOPING ELECTRONIC COUNTERMEASURES AGAINST SURFACE-TO-AIR MISSILE ATTACKS.

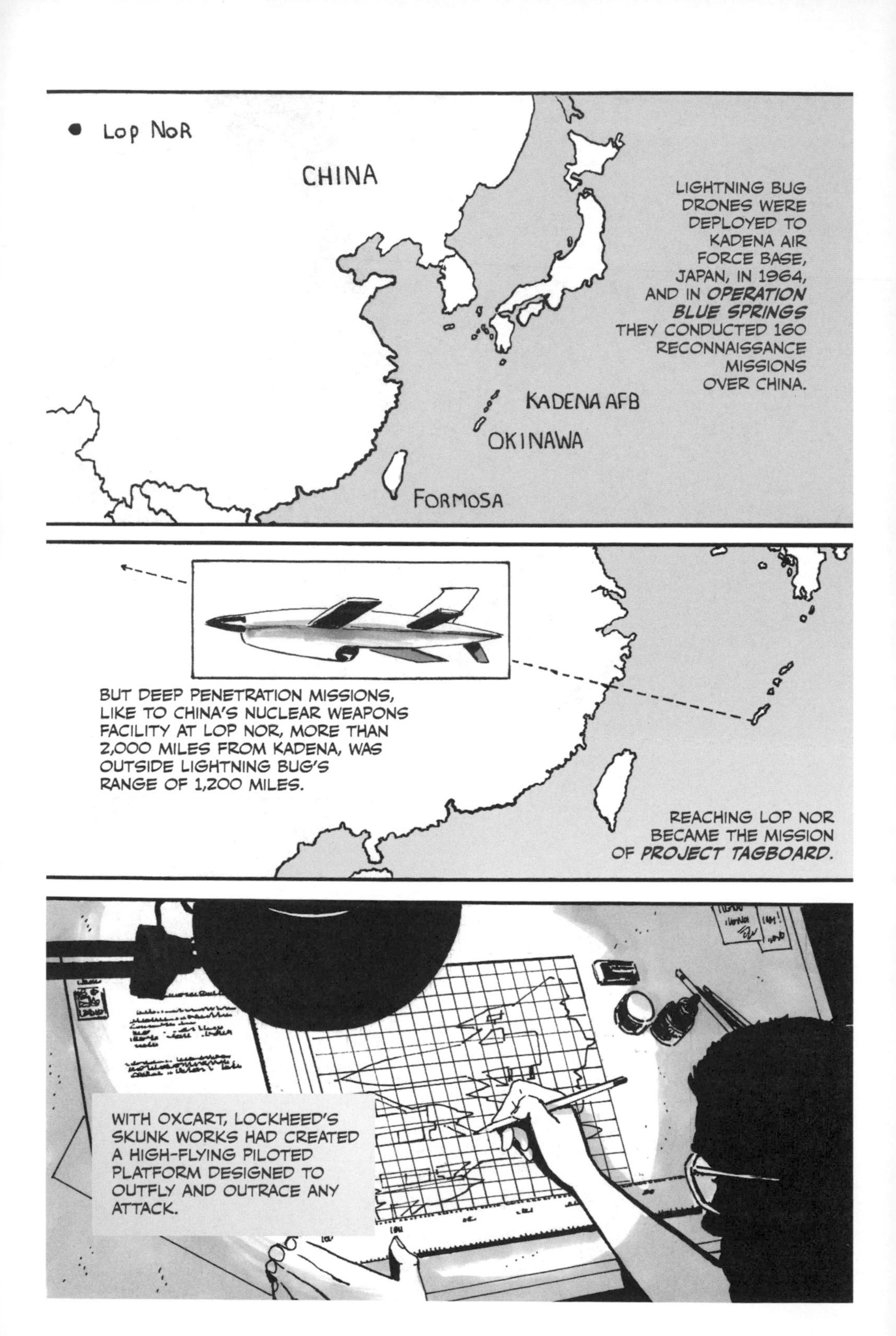

Lop NoR
CHINA
LIGHTNING BUG DRONES WERE DEPLOYED TO KADENA AIR FORCE BASE, JAPAN, IN 1964, AND IN *OPERATION BLUE SPRINGS* THEY CONDUCTED 160 RECONNAISSANCE MISSIONS OVER CHINA.
KADENA AFB
OKINAWA
FORMOSA
BUT DEEP PENETRATION MISSIONS, LIKE TO CHINA'S NUCLEAR WEAPONS FACILITY AT LOP NOR, MORE THAN 2,000 MILES FROM KADENA, WAS OUTSIDE LIGHTNING BUG'S RANGE OF 1,200 MILES.
REACHING LOP NOR BECAME THE MISSION OF *PROJECT TAGBOARD*.
WITH OXCART, LOCKHEED'S SKUNK WORKS HAD CREATED A HIGH-FLYING PILOTED PLATFORM DESIGNED TO OUTFLY AND OUTRACE ANY ATTACK.

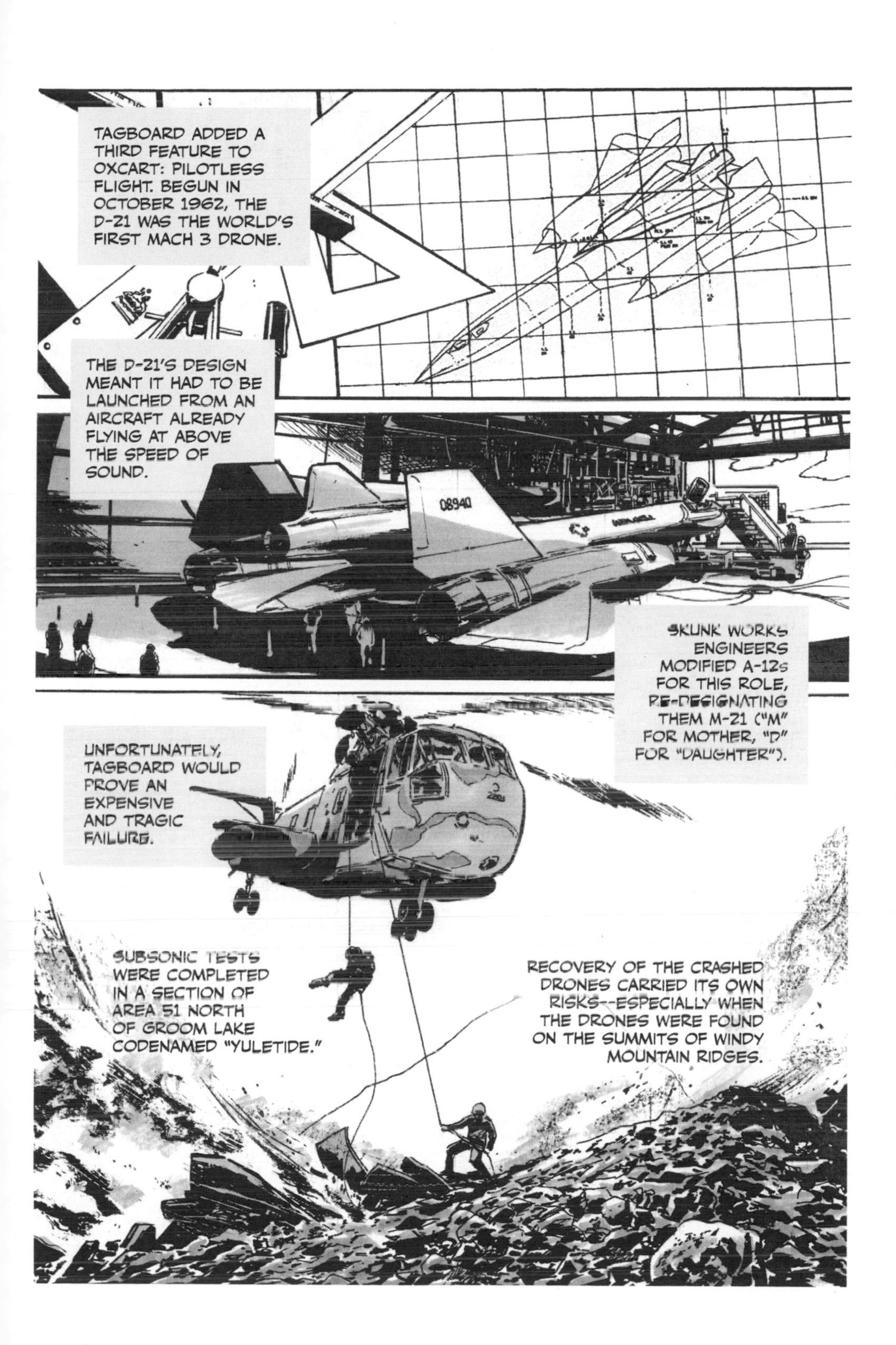
TAGBOARD ADDED A THIRD FEATURE TO OXCART: PILOTLESS FLIGHT. BEGUN IN OCTOBER 1962, THE D-21 WAS THE WORLD'S FIRST MACH 3 DRONE.
THE D-21'S DESIGN MEANT IT HAD TO BE LAUNCHED FROM AN AIRCRAFT ALREADY FLYING AT ABOVE THE SPEED OF SOUND.
08940
SKUNK WORKS ENGINEERS MODIFIED A-12s FOR THIS ROLE, RE-DESIGNATING THEM M-21 ("M" FOR MOTHER, "D" FOR "DAUGHTER").
UNFORTUNATELY, TAGBOARD WOULD PROVE AN EXPENSIVE AND TRAGIC FAILURE.
SUBSONIC TESTS WERE COMPLETED IN A SECTION OF AREA 51 NORTH OF GROOM LAKE CODENAMED "YULETIDE."
RECOVERY OF THE CRASHED DRONES CARRIED ITS OWN RISKS--ESPECIALLY WHEN THE DRONES WERE FOUND ON THE SUMMITS OF WINDY MOUNTAIN RIDGES.

FINALLY, TAGBOARD'S FIRST MACH 3 LAUNCH WAS SCHEDULED: THE NIGHT OF JULY 30, 1966, OFF THE COAST OF CALIFORNIA.
BILL PARK WAS THE PILOT AND RAY TORICK WAS HIS FLIGHT ENGINEER.
COLONEL HUGH "SLIP" SLATER, THE "MAYOR" OF AREA 51 SINCE 1965, WOULD FOLLOW AND FILM THE TEST FLIGHT IN AN A-12. HE WAS WORRIED.
BILL, IT'S A DANGEROUS MISSION. ONLY A FEW FEET SEPARATE THE D-21 FROM THE TAIL OF YOUR M-21.
I'LL BE OKAY.

BUT WHEN PARK'S PLANE REACHED THE LAUNCH POINT 150 MILES OFF THE CALIFORNIA COAST, DISASTER STRUCK.

INSTEAD OF CONTINUING UP AFTER SEPARATION, THE RELEASED DRONE FLEW DOWN.
MIRACULOUSLY, PARKS AND TORICK WERE NOT HIT BY THE DRONE, SUCCESSFULLY EJECTED, AND MANAGED TO AVOID THE WRECKAGE.
THE MISSION'S RESCUE SHIP SUCCESSFULLY PULLED PARK OUT OF THE OCEAN.
UNFORTUNATELY, TORICK HAD BECOME ENTANGLED IN HIS PARACHUTE AFTER LANDING...
I'M SORRY, SIR. HE'S DEAD.

KELLY JOHNSON WAS DEVASTATED.
I WILL NOT RISK ANY MORE TEST PILOTS OR A-12s. I DON'T HAVE EITHER TO SPARE.
HE RETURNED ALL TAGBOARD FUNDING TO THE AIR FORCE AND CIA.
BUT INSTEAD OF SHUTTING DOWN TAGBOARD, THE AIR FORCE REVISED IT SO DRONES EQUIPPED WITH ROCKET BOOSTERS COULD BE LAUNCHED BENEATH THE WING OF A SUBSONIC B-52, LIKE IT HAD DONE WITH THE X-15.
UNDER THE CODENAME SENIOR BOWL, FOUR MISSIONS WERE FLOWN, CONDUCTING RECONNAISSANCE OVER CHINA'S LOP NOR NUCLEAR TEST SITE FROM NOVEMBER 9, 1969, TO MARCH 20, 1971. ALL WERE FAILURES, WITH THE DRONES BEING LOST AND NO FILM RECOVERED.
BIRD DOG TO KENNEL. MID-AIR RECOVERY FAILED. PELICAN IS SINKING. DID YOU CATCH ITS FISH?
KENNEL TO BIRD DOG, NO JOY HERE. PELICAN'S FISH JOINING IT BELOW.

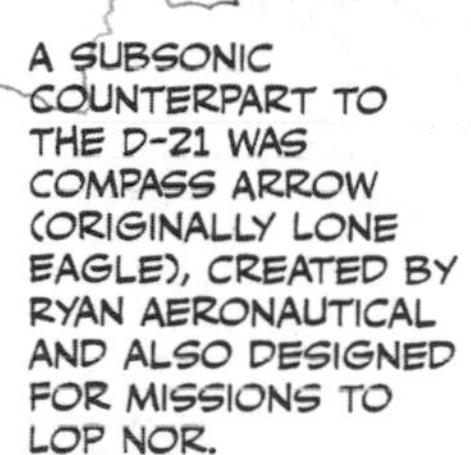
A SUBSONIC COUNTERPART TO THE D-21 WAS COMPASS ARROW (ORIGINALLY LONE EAGLE), CREATED BY RYAN AERONAUTICAL AND ALSO DESIGNED FOR MISSIONS TO LOP NOR.

BEGUN IN THE MID-1960s, COMPASS ARROW WENT WILDLY OVER BUDGET. THE 10 DRONES BUILT HAD A UNIT COST OF $65 MILLION.
WORSE, BY THE TIME IT WAS OPERATIONAL IN THE EARLY 1970s, IT WAS OBSOLETE.
THE TEST FLIGHT TODAY CALLS FOR A FLIGHT TO TARGET OF 100 YARDS.
I'LL SET UP THE TARGET.
AREA 51 WAS ALSO THE TEST SITE FOR WHAT WERE IN THE 1970s THE WORLD'S SMALLEST DRONES, IN PROJECT INSECTOTHOPTER.
I'VE GOT THE LASER GUIDANCE SYSTEM.
ALL RIGHT. THEN ...
... LET'S GO!
RESEMBLING A DRAGONFLY, THE GAS-POWERED, MINIATURE-CAMERA-EQUIPPED INSECTOTHOPTER WAS GUIDED TO ITS TARGET BY A LASER BEAM. UNFORTUNATELY, LIKE ITS REAL-WORLD COUNTERPART ...
DAMN THAT CROSSWIND!
... IT WAS INCAPABLE OF FLYING AGAINST THE WIND.

WITH PROJECT AQUILINE, THE CIA AGAIN USED A MEMBER OF THE ANIMAL KINGDOM AS COVER, IN THIS CASE, A CONDOR.
GET READY TO LAUNCH!

POWERED BY A LAWNMOWER ENGINE, IT WAS REMOTELY GUIDED BY A VIDEO LINK.

VIDEO LINK IS GOOD. FLYING SMOOTH AS SILK.

BUT HIGH COST AND IMMATURE MINIATURIZATION TECHNOLOGY TURNED THE AQUILINE CONDOR INTO AN ALBATROSS.

LIKE SO MANY OTHER DRONE PROJECTS DEVELOPED AND THEN TESTED AT AREA 51 IN THE 1960s AND 1970s, AQUILINE WAS CANCELED.
IT WOULD BE YEARS BEFORE TECHNOLOGY ADVANCED ENOUGH TO MAKE DRONES PRACTICAL FOR THE BATTLEFIELD.

CHAPTER FOUR
GOING UNDERGROUND

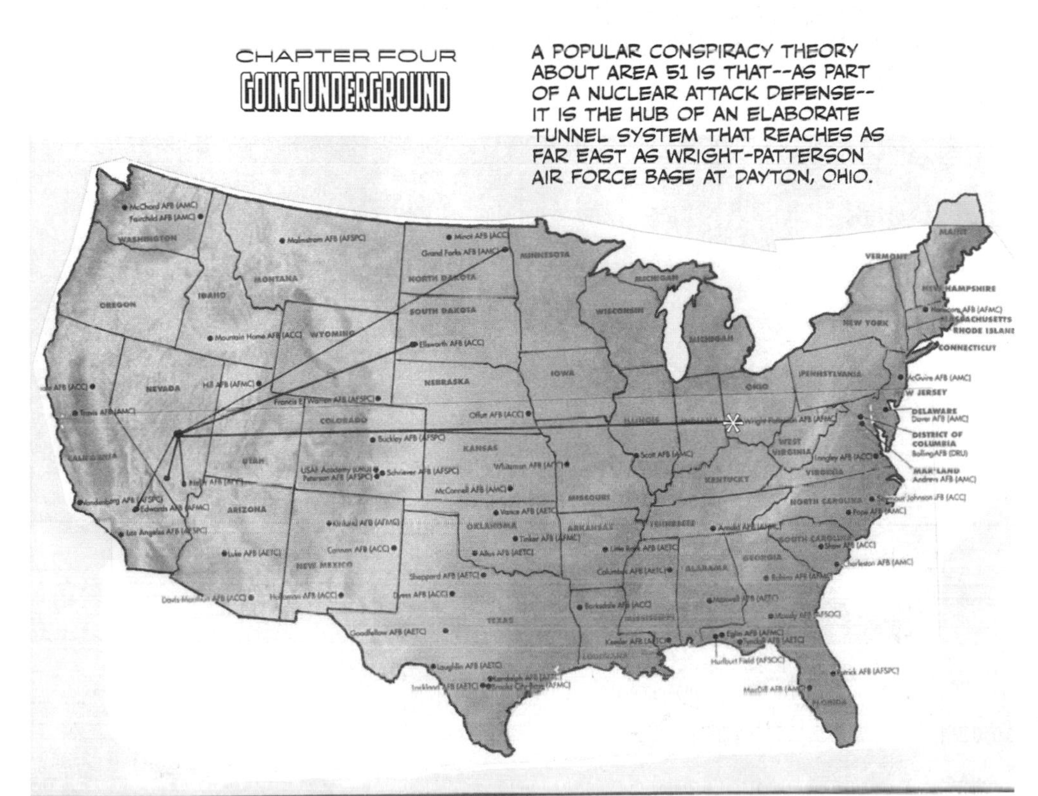

AT THE HEIGHT OF THE SPACE RACE, AND IN ORDER TO AVERT CHAOS, IN 1967 THE UNITED STATES, THE UNITED KINGDOM, AND THE SOVIET UNION SIGNED THE *OUTER SPACE TREATY*...
UK
USA

... AN AGREEMENT THAT PROHIBITED THE USE OF OUTER SPACE OR THE MOON AS LAUNCH PLATFORMS FOR WEAPONS OF MASS DESTRUCTION.

BY 2011, MORE THAN 8,000 SATELLITES OF ALL TYPES WOULD BE ORBITING EARTH.
BUT EVEN BEFORE SPY SATELLITES BECAME COMMON, THE UNITED STATES AND OTHER COUNTRIES WERE BUILDING TUNNEL SYSTEMS TO HIDE AND PROTECT THEIR SENSITIVE FACILITIES FROM EYES IN THE SKY.

AREA 51 BECAME HOME TO A VARIETY OF SYSTEMS DESIGNED TO *FIND AND DESTROY* UNDERGROUND FACILITIES.
DETECTION SYSTEMS INCLUDED GROUND-PENETRATING RADAR.
I FEEL LIKE A WATER WIZARD WITH A HIGH TECH DOWSING ROD.
OR A BEACH BUM WITH A FANCY METAL DETECTOR.
THAT'S A PRETTY GOOD READ OF THE TUNNEL. WHAT'S ITS DEPTH?
A HUNDRED FEET.
OKAY. LET'S GO OVER TO THE NEXT GRID AND SEE WHAT WE FIND AT 500 FEET.

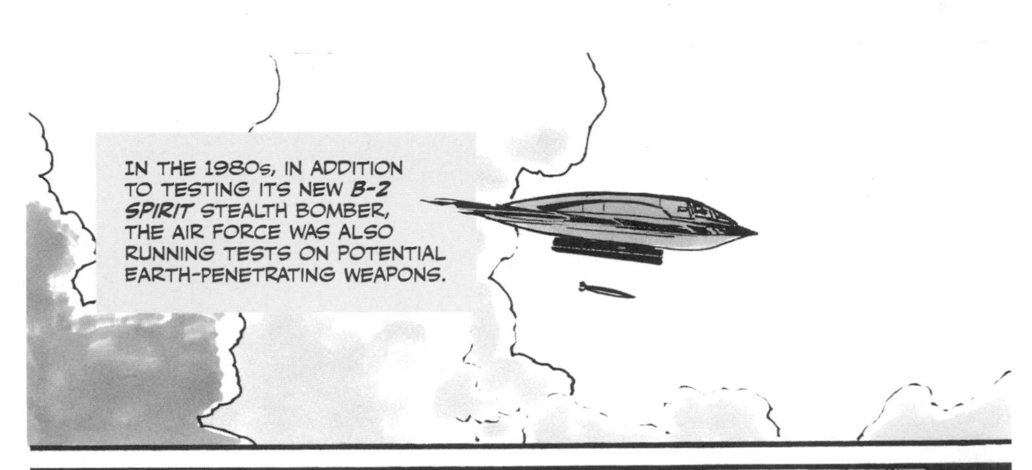
IN THE 1980s, IN ADDITION TO TESTING ITS NEW *B-2 SPIRIT* STEALTH BOMBER, THE AIR FORCE WAS ALSO RUNNING TESTS ON POTENTIAL EARTH-PENETRATING WEAPONS.

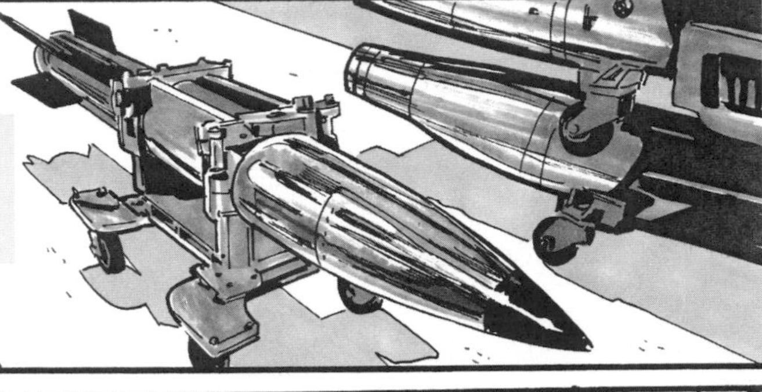
THE WARHEAD USED WAS THE *B61-11*, THE LATEST MODIFICATION OF THE B61 NUCLEAR BOMB.

CONFIGURED TO CARRY A 10-MEGATON NUCLEAR WARHEAD, THE BOMB WAS DESIGNED TO PENETRATE THE GROUND ...

K-BOOM
... BEFORE EXPLODING, SENDING OUT A SHOCK WAVE THAT WOULD DESTROY A TUNNEL.

ANALYSIS OF NON-NUCLEAR TEST DROPS FROM 40,000 FEET REVEALED THAT THE BOMB COULD NOT PENETRATE DEEP ENOUGH TO ACCOMPLISH THE *DUAL GOALS* OF TARGET DESTRUCTION AND MINIMAL RADIOACTIVE FALLOUT.
WE NEEDED TO PENETRATE, WHAT--850 FEET?
YES, AND MAXIMUM PENETRATION WAS NOT EVEN 100 FEET IN DIRT--
--AND LESS THAN 30 FEET IN SOLID ROCK!
WHAT IF WE CHANGED THE CASING TO SOMETHING HARDER?
NON-STARTER.
THE MOST PENETRATION IN GRANITE, UNDER WHICH THE TARGETED FACILITIES ARE BURIED, WITH THE STRONGEST MATERIAL AVAILABLE IS MAYBE 100 FEET--
--AND THAT'S STRETCHING IT.
OKAY, TEAM. THEN IT'S BACK TO THE DRAWING BOARD.
UNDER THE GEORGE W. BUSH ADMINISTRATION, STUDIES CONTINUED UNDER THE *ROBUST NUCLEAR EARTH PENETRATOR PROGRAM*.
BUT CONGRESS CUT FUNDING FOR IT IN 2005, AND THE PROGRAM WAS ENDED.

THE NON-NUCLEAR OPTION THE AIR FORCE FOCUSED ON WAS THE USE OF KINETIC ENERGY WEAPONS.
THE CONCEPT WAS ORIGINALLY PROPOSED UNDER THE NAME *PROJECT THOR* BY SCIENCE FICTION WRITER JERRY POURNELLE IN THE 1950S WHEN HE WAS WORKING FOR BOEING.
ONE PROPOSED SYSTEM INVOLVED LAUNCHING INERT PROJECTILES, IN THIS CASE TUNGSTEN RODS A FOOT IN DIAMETER AND UP TO 20 FEET IN LENGTH, FROM AN ORBITING SATELLITE.
THE GUIDED *"RODS FROM GOD,"* AS THEY CAME TO BE KNOWN, WERE RATED TO REACH A SPEED OF 36,000 FEET PER SECOND ...

... THE SAME SPEED AS THE METEOR THAT STRUCK *CHELYABINSK, SIBERIA,* IN MARCH 2013--WHICH RELEASED 20 TO 30 TIMES MORE ENERGY THAN THE ATOMIC BOMB THAT DESTROYED HIROSHIMA.

OTHER *KEW* SYSTEMS INCLUDE PROJECTILES ELECTROMAGNETICALLY LAUNCHED FROM A RAIL GUN...

...AND FROM A BALLISTIC MISSILE.

A FOURTH OPTION BEING STUDIED INVOLVES THE USE OF HOUSE-SIZED METEORS INTERCEPTED IN SPACE AND GUIDED TO TARGETS ON EARTH.

THE AIR FORCE HAS REFUSED TO COMMENT HOW FAR ALONG ANY OF THESE PROPOSED SYSTEMS ARE.

CHAPTER FIVE
FROM HOPELESS DIAMOND TO HAVE BLUE TO F-117A NIGHTHAWK
ON JANUARY 17, 1975, LOCKHEED'S ADVANCED DEVELOPMENT PROJECT (SKUNK WORKS' OFFICIAL NAME), GOT A NEW BOSS. BEN RICH BECAME ITS VICE PRESIDENT FOR ADVANCED PROJECTS. HE WAS HANDPICKED BY RETIRING KELLY JOHNSON TO BE HIS SUCCESSOR.

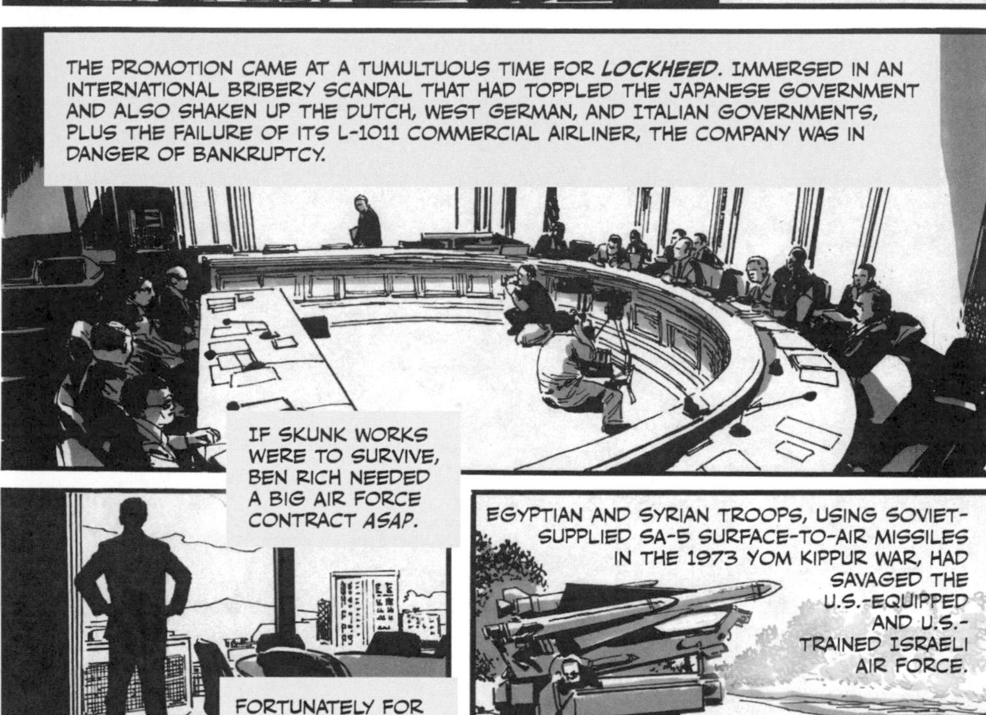
THE PROMOTION CAME AT A TUMULTUOUS TIME FOR LOCKHEED. IMMERSED IN AN INTERNATIONAL BRIBERY SCANDAL THAT HAD TOPPLED THE JAPANESE GOVERNMENT AND ALSO SHAKEN UP THE DUTCH, WEST GERMAN, AND ITALIAN GOVERNMENTS, PLUS THE FAILURE OF ITS L-1011 COMMERCIAL AIRLINER, THE COMPANY WAS IN DANGER OF BANKRUPTCY.
IF SKUNK WORKS WERE TO SURVIVE, BEN RICH NEEDED A BIG AIR FORCE CONTRACT ASAP.
FORTUNATELY FOR HIM, THE AIR FORCE HAD A NEED, TOO.
EGYPTIAN AND SYRIAN TROOPS, USING SOVIET-SUPPLIED SA-5 SURFACE-TO-AIR MISSILES IN THE 1973 YOM KIPPUR WAR, HAD SAVAGED THE U.S.-EQUIPPED AND U.S.-TRAINED ISRAELI AIR FORCE.

SENIOR PENTAGON OFFICIALS FEARED A SIMILAR FATE FOR THEIR AIRCRAFT IN ANY POTENTIAL CONFLICT WITH SOVIET UNION AND WARSAW PACT FORCES.

A TECHNICAL PAPER BY ***PYOTR UFIMTSEV,*** CHIEF SCIENTIST AT MOSCOW'S INSTITUTE OF RADIO ENGINEERING, WAS JUST TRANSLATED.

IT'S CALLED *"METHOD OF EDGE WAVES IN THE PHYSICAL THEORY OF DIFFRACTION"* AND IT IS AS OBTUSE AND IMPENETRABLE AS YOU'D EXPECT ...

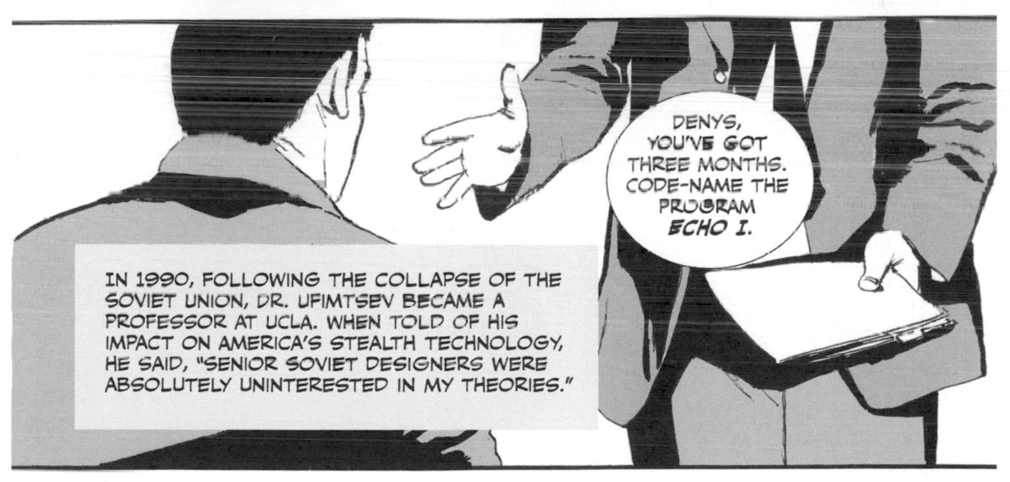

FIVE WEEKS LATER ON MAY 5, 1975...
BOSS, MEET...

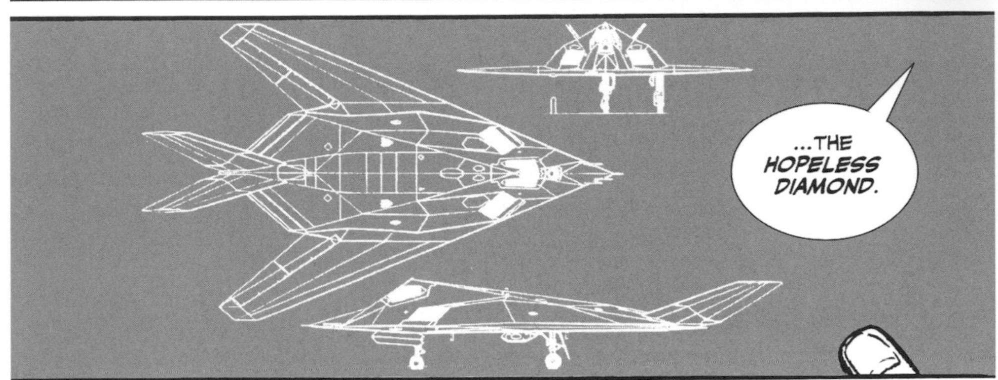
...THE *HOPELESS DIAMOND.*

THIS SHAPE IS ONE THOUSAND TIMES LESS VISIBLE ON RADAR THAN THE LEAST VISIBLE SHAPE PREVIOUSLY PRODUCED AT THE SKUNK WORKS.
ON A RADAR SCREEN IT WOULD APPEAR AS A...WHAT? AS BIG AS A *CONDOR*, AN EAGLE, AN OWL--?

BEN, TRY AS BIG AS AN EAGLE'S EYEBALL!

THE DESIGN WASN'T JUST RADICAL, IT VIOLATED THE LAW OF AERODYNAMICS. WHEN KELLY JOHNSON, NOW A CONSULTANT, SAW THE SKETCHES, HIS FAMOUS TEMPER ERUPTED.
BEN, HAVE YOU LOST YOUR MIND? THIS CRAP WILL NEVER GET OFF THE GROUND!
BUT RICH STUCK TO HIS GUNS. PRELIMINARY TESTS ON A HOPELESS DIAMOND MODEL CONFIRMED THE THEORETICAL DATA.
IN APRIL 1976, LOCKHEED RECEIVED THE CONTRACT TO BUILD TWO EXPERIMENTAL AIRCRAFT IN A PROGRAM CODENAMED *HAVE BLUE*.
AS CONSTRUCTION OF THE HAVE BLUE PROTOTYPES COMMENCED, RICH BEGAN HEARING RUMORS THAT THE SKUNK WORKS HAD RIGGED THE RADAR TEST RESULTS.
FURIOUS, RICH ASKED RADAR EXPERT MIT PROFESSOR *LINDSAY ANDERSON* TO CONDUCT INDEPENDENT TESTS.
BALL BEARINGS RANGING FROM AN EIGHTH OF AN INCH IN SIZE TO THAT OF A GOLF BALL WERE SUCCESSIVELY MOUNTED TO THE NOSE OF THE HOPELESS DIAMOND MODEL FOR ANDERSON TO TAKE HIS RADAR READINGS.

IN HIS NEXT VISIT TO THE PENTAGON, IN ADDITION TO THE TEST RESULTS, RICH BROUGHT WITH HIM SOMETHING DESIGNED TO UNDERSCORE THEM.
GENERALS...
...HERE'S THE OBSERVABILITY OF THE HAVE BLUE ON RADAR.
MAJOR, HERE'S YOUR NIGHTHAWK. READY TO TAKE YOUR ACCEPTANCE FLIGHT?
PRODUCTION FOR THE STEALTH FIGHTER, NOW DESIGNATED THE *F-117A NIGHTHAWK*, BEGAN IN NOVEMBER 1978.
BY 1982, OPERATIONAL MODELS WERE WAITING FOR THEIR AIR FORCE PILOTS ON THE FIELD AT AREA 51.
I'VE DONE MY TIME IN SIMULATORS. LET'S DO THE REAL THING.
I'VE SEEN THE STEALTH DATA. IT'S IMPRESSIVE, BUT IS IT *REALLY* THAT GOOD?
YEP.
THAT'S A PRETTY CONFIDENT-SOUNDING "YEP."
WHAT MAKES YOU SO COCK-SURE?
YEP.
THEM.
IF A BAT'S RADAR CAN'T GET A HIT OFF A NIGHTHAWK, HOW MUCH CHANCE DO YOU GIVE RUSSIAN RADAR?

BUT AS THE FIRST AIR FORCE PILOTS WERE GETTING FAMILIAR WITH THEIR REVOLUTIONARY BOMBER, CIRCA 1982, AN EVENT OCCURRED THAT ALMOST MADE AREA 51 AND EVERYTHING HAPPENING IN IT HEADLINE NEWS.
WE'RE UNDER ATTACK!
NEWS OF A HELICOPTER ATTACK ON AREA 51 ROCKETED ALL THE WAY UP TO THE WHITE HOUSE.
SQUADRONS AT NEARBY NELLIS AIR FORCE BASE WERE BEING SCRAMBLED...
...AND NUCLEAR SUBMARINES WERE PUT ON ALERT, WITH TOMAHAWK NUCLEAR MISSILES BEING PROGRAMMED TO STRIKE AREA 51.
THEN, JUST AS SUDDENLY...
STAND DOWN. MISSION IS CANCELED. REPEAT. MISSION IS CANCELED.

INCREDIBLY, THE PRIVATE CONTRACTOR RESPONSIBLE FOR SECURITY AT AREA 51 HAD NEGLECTED TO INFORM EVERYONE THEY WERE CONDUCTING A MOCK ATTACK TO TEST THEIR SECURITY SYSTEMS.
AREA 51--AND EVERYTHING ON IT--HAD COME WITHIN A HAIRSBREADTH OF BEING NOT ONLY SPECTACULARLY DECLASSIFIED, BUT ALSO ANNIHILATED.

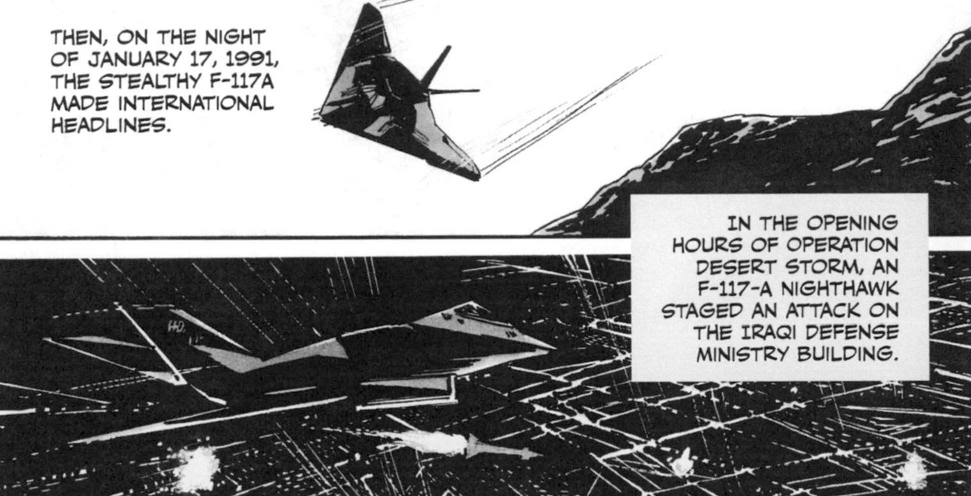
THEN, ON THE NIGHT OF JANUARY 17, 1991, THE STEALTHY F-117A MADE INTERNATIONAL HEADLINES.
IN THE OPENING HOURS OF OPERATION DESERT STORM, AN F-117-A NIGHTHAWK STAGED AN ATTACK ON THE IRAQI DEFENSE MINISTRY BUILDING.

THE PENTAGON RELEASED A VIDEO OF THE ATTACK--SUDDEN, SWIFT, AND DEVASTATING--THAT BECAME AN INTERNATIONAL SENSATION.

CHAPTER SIX

BEWARE THE PREDATOR

"MOORE'S LAW," AS IT CAME TO BE KNOWN, ESTABLISHED THAT EVERY TWO YEARS, THE POWER OF THE NEXT-GENERATION MICROCHIPS DOUBLES.

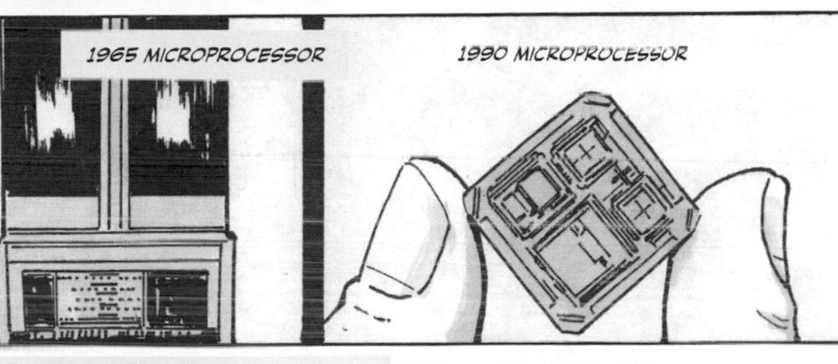

... THE *PREDATOR DRONE.*
BUT THE PREDATOR'S MOST PASSIONATE ADVOCATE WAS NOT THE CIA ...

... OR THE ARMY, WHO WOULD HAVE APPEARED TO BE THE MOST LIKELY SUPPORTERS.

INSTEAD, IT WAS THE *UNITED STATES AIR FORCE.*

THERE'S A CERTAIN IRONY IN THAT. AFTER ALL, THE AIR FORCE ETHOS IS CENTERED ON PILOTED AIRCRAFT, WITH FIGHTER PILOTS BEING AT THE TOP OF AIR FORCE HIERARCHY.
A SUCCESSFUL PILOTLESS PREDATOR WOULD HAVE THE SAME IMPACT AS THE SUBMARINE
... AND THE AIRCRAFT CARRIER HAD ...
... IN USHERING THE DECLINE IN WORLD WAR II OF THE "BIG GUN" NAVY, WHICH WAS FORMERLY CENTERED ON THE BATTLESHIP.
DRONES COULD DRAMATICALLY REDUCE THE NEED FOR THE USE OF PILOTED JET FIGHTERS IN A WAR ZONE.

IN 1982, THE DEFENSE ADVANCED RESEARCH PROJECTS AGENCY (DARPA), RESPONSIBLE FOR DEVELOPING NEW TECHNOLOGIES FOR THE MILITARY, IDENTIFIED THE NEED FOR AN INEXPENSIVE, RELIABLE UNMANNED AERIAL VEHICLE (UAV) WITH LONG-ENDURANCE RECONNAISSANCE CAPABILITY.

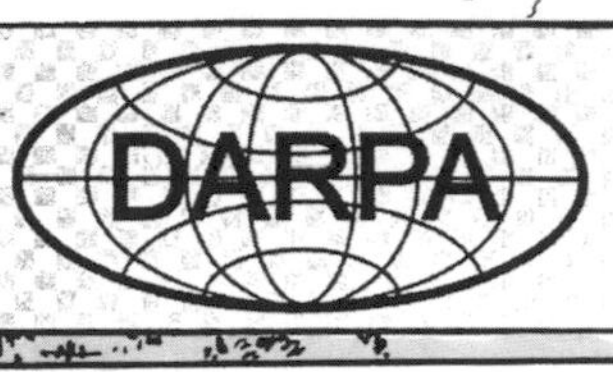

THE WINNER OF THE DESIGN COMPETITION WAS ***ABRAHAM KAREM***, AN ISRAELI ÉMIGRÉ WHO HAD DESIGNED HIGH-TECH WEAPONRY FOR THE ISRAELI DEFENSE FORCES. HIS ENTRY WAS THE ***AMBER***, THE GRANDFATHER OF THE PREDATOR.

WHEN FURTHER DEVELOPMENT ON AMBER STALLED, KAREM BUILT A CHEAPER, LOWER-TECH VERSION CALLED THE ***GNAT-750***.

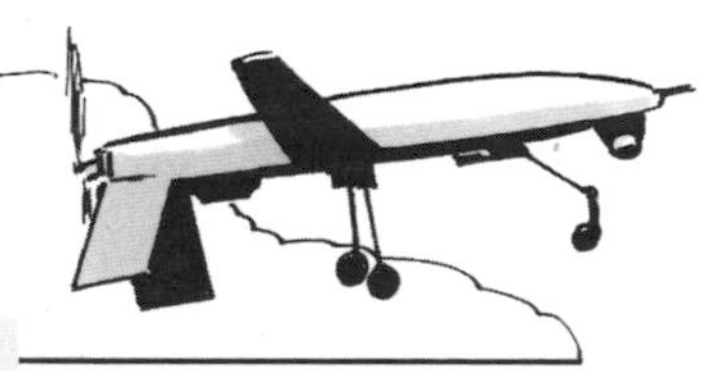

WITH THE SITUATION BECOMING DIRE IN THE BALKANS FOLLOWING THE BREAKUP OF YUGOSLAVIA IN 1992, THE CIA PURCHASED TWO GNAT-750s TO OBSERVE EVENTS IN BOSNIA.

MEANWHILE, KAREM WAS USING THE EXPERIENCE HE GAINED ON AMBER AND GNAT-750 TO CREATE THE RECONNAISSANCE ***RQ-1 PREDATOR***.

GENERAL RONALD R. FOGLEMAN HAD BEEN FOLLOWING THE DEVELOPMENT OF THE RQ-1 PREDATOR AND SAW ITS POTENTIAL. WHEN HE BECAME AIR FORCE CHIEF OF STAFF IN OCTOBER 1994, HE ACTED.

HE ACTIVATED THE FIRST PREDATOR SQUADRON, THE ***11TH RECONNAISSANCE SQUADRON,*** ON JULY 29, 1995.

THE PREDATOR'S SUCCESS INSPIRED THE NEXT STEP, TURNING THE UAV PREDATOR INTO A *UCAV*--UNMANNED COMBAT AERIAL VEHICLE--ARMED WITH A MODIFIED HELLFIRE MISSILE.
THE ***HELLFIRE*** WAS LIGHT *(UNDER 100 POUNDS)* AND FAST *(950 MPH)*, AND IT HAD RANGE *(8,750 YARDS)* AND PUNCH *(18 LB. HIGH EXPLOSIVE WARHEAD)*.
BUT, UNLIKE PREVIOUS WEAPONS SYSTEMS TESTED AT AREA 51 ...
... THE UCAV PREDATOR--DESIGNATED MQ-1 ("M" FOR MULTIROLE)--WAS CREATED TO TARGET INDIVIDUALS.
WHOOM!

OSAMA BIN LADEN AND THE LEADERSHIP OF HIS TERRORIST ORGANIZATION, AL QAEDA.

ON AUGUST 7, 1998, AL QAEDA, A HERETOFORE LITTLE-KNOWN TERRORIST GROUP LED BY EXPATRIATE SAUDI NATIONAL OSAMA BIN LADEN, CONDUCTED TRUCK-BOMB ATTACKS AGAINST U.S. EMBASSIES IN DAR ES SALAAM AND NAIROBI, KILLING 224 PEOPLE AND WOUNDING 4,000.
AND, ON OCTOBER 12, 2000, A GROUP OF ITS MEMBERS ATTACKED THE USS *COLE*, DOCKED IN THE YEMENI PORT OF ADEN, KILLING 17 AMERICAN SAILORS AND WOUNDING 39.
THEN, ON ***SEPTEMBER 11, 2001,*** AL QAEDA STAGED THE GREATEST TERRORIST ATTACK IN HISTORY, SUCCESSFULLY CRASHING TWO COMMERCIAL AIRLINERS INTO THE TWIN TOWERS OF THE WORLD TRADE CENTER AND ONE INTO THE PENTAGON.

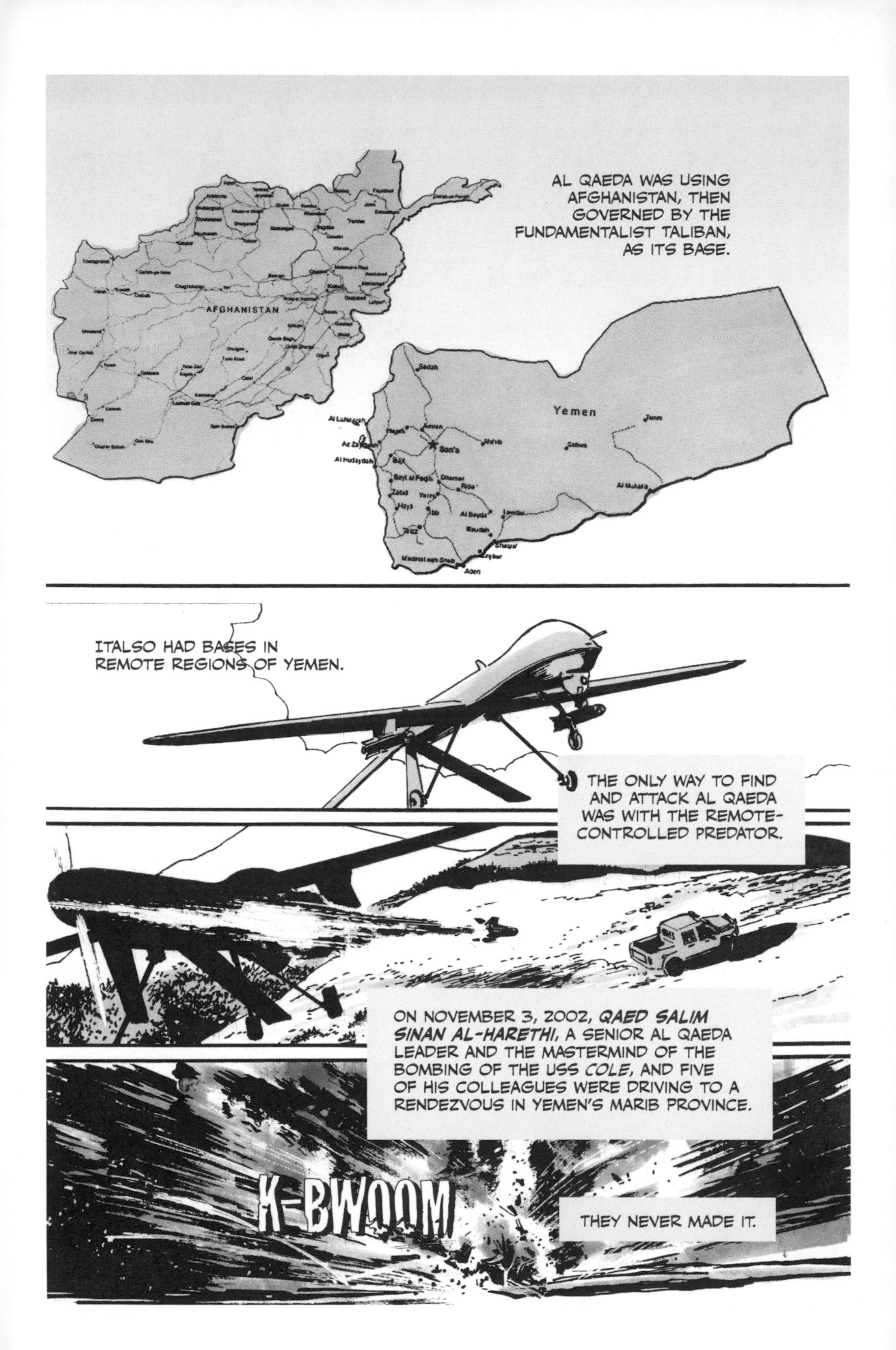
AL QAEDA WAS USING AFGHANISTAN, THEN GOVERNED BY THE FUNDAMENTALIST TALIBAN, AS ITS BASE.
AFGHANISTAN
Yemen
ITALSO HAD BASES IN REMOTE REGIONS OF YEMEN.
THE ONLY WAY TO FIND AND ATTACK AL QAEDA WAS WITH THE REMOTE-CONTROLLED PREDATOR.
ON NOVEMBER 3, 2002, *QAED SALIM SINAN AL-HARETHI*, A SENIOR AL QAEDA LEADER AND THE MASTERMIND OF THE BOMBING OF THE USS *COLE*, AND FIVE OF HIS COLLEAGUES WERE DRIVING TO A RENDEZVOUS IN YEMEN'S MARIB PROVINCE.
K-BWOOM
THEY NEVER MADE IT.

THAT ATTACK MADE THE CLASSIFIED PREDATOR WORLD FAMOUS AND DRAMATICALLY ANNOUNCED A CHANGE IN HOW THE UNITED STATES, AND OTHER MAJOR POWERS, WOULD WAGE WAR.
SOON, THE PREDATOR WAS JOINED WITH THE LARGER *MQ-9 REAPER*, CAPABLE OF CARRYING HEAVIER GBU-12 PAVEWAY II LASER-GUIDED BOMBS.
AND THE *RQ-4 GLOBAL HAWK* RECONNAISSANCE DRONE.

CHAPTER SEVEN

TEST BED FOR TECHNOLOGIES, NOT OPERATIONS

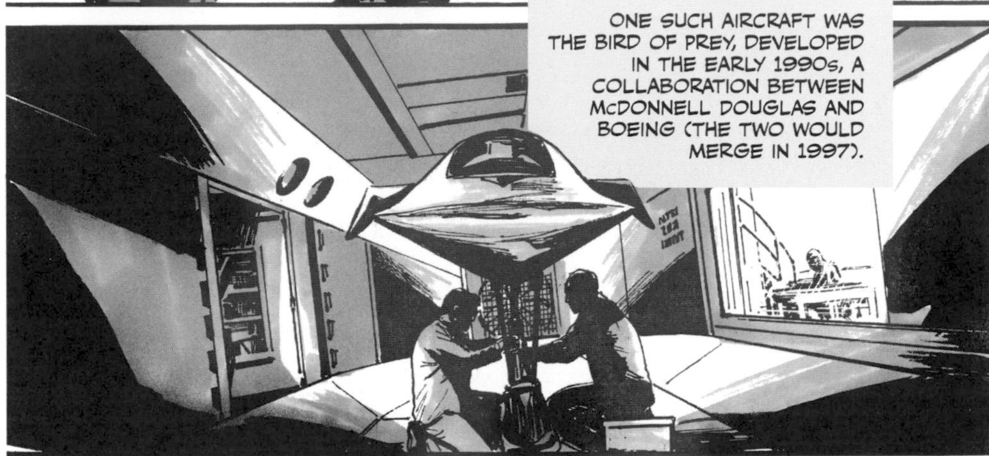

I'M GOING TO FOLD THIS PIECE OF PAPER INTO AN AIRPLANCE WITH INVERTED WINGTIPS LIKE YOUR BIRD OF PREY.
NOW WATCH!
WELL?
JIM, OUR COMPUTER SIMULATIONS AND WIND TUNNEL TESTS INDICATE THAT WON'T HAPPEN.
WE RAN THE TESTS DOZENS OF TIMES-- THEY ALL CAME UP GOOD.
THEY ALWAYS DO-- IN THEORY.

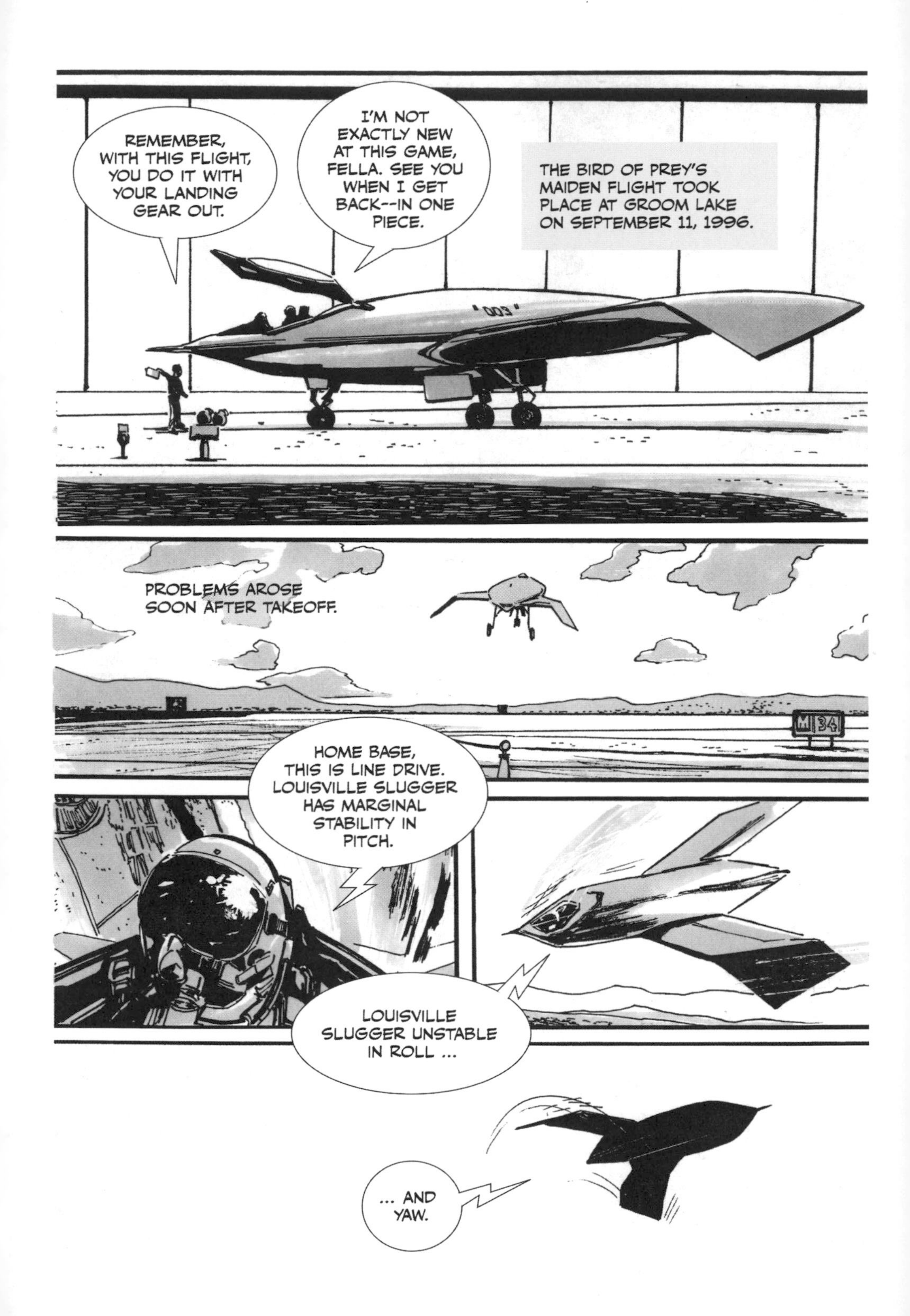
REMEMBER, WITH THIS FLIGHT, YOU DO IT WITH YOUR LANDING GEAR OUT.
I'M NOT EXACTLY NEW AT THIS GAME, FELLA. SEE YOU WHEN I GET BACK--IN ONE PIECE.
THE BIRD OF PREY'S MAIDEN FLIGHT TOOK PLACE AT GROOM LAKE ON SEPTEMBER 11, 1996.
"003"
PROBLEMS AROSE SOON AFTER TAKEOFF.
M 34
HOME BASE, THIS IS LINE DRIVE. LOUISVILLE SLUGGER HAS MARGINAL STABILITY IN PITCH.
LOUISVILLE SLUGGER UNSTABLE IN ROLL ...
... AND YAW.

THE TEST FLIGHT REVEALED OTHER MANEUVERABILITY PROBLEMS.
THOSE PROBLEMS SHOULDN'T HAVE HAPPENED!
YEAH, REALITY CAN BE LIKE THAT. BUT ANY FLIGHT YOU CAN WALK AWAY FROM IS A GOOD FLIGHT.
DON'T WORRY, WE'LL FIND AND FIX THE PROBLEMS FOR THE NEXT FLIGHT.
YEAH, YOU DO THAT. YOU'LL FIND ME AT THE HOUSE SIX BAR.
BY ITS FIFTH FLIGHT, THE MANEUVERABILITY PROBLEMS HAD BEEN SOLVED ENOUGH TO ALLOW FOR ADDITIONAL TESTS WITH THE LANDING GEAR RETRACTED.
LESSONS LEARNED FROM TECHNOLOGY AND DESIGN IN THE BIRD OF PREY WERE LATER USED IN A VARIETY OF BOEING AIRCRAFT, INCLUDING THE X-45A UNMANNED COMBAT AIR VEHICLE ...
... AND THE X-32, BOEING'S ENTRY IN THE JOINT STRIKE FIGHTER (JSF) COMPETITION.

CHAPTER EIGHT

FLIGHTS OR FANTASIES?

DECEPTION AND DISSEMBLING HAVE BEEN AN INTEGRAL PART OF ACTIVITY AT AREA 51 FROM ITS INCEPTION.

OF ALL THE TOP-SECRET AIRCRAFT SLATED FOR TESTING AT AREA 51, SOME NEVER GOT PAST THE DRAWING BOARD. OTHERS FAILED FOR A VARIETY OF REASONS IN THE PROTOTYPE STAGE. ONLY A FEW BECAME OPERATIONAL.

THEN THERE WERE THOSE THAT MAY HAVE BEEN COMPLETE FABRICATIONS WHOSE ULTIMATE PURPOSE WAS, AND REMAINS...UNKNOWN. UNCONFIRMED AND UNVERIFIED, WE NOW EXPLORE THE POSSIBLE ***FANCIFUL FRAUDS*** THAT ARE A PART OF THE ***LORE*** OF AREA 51.

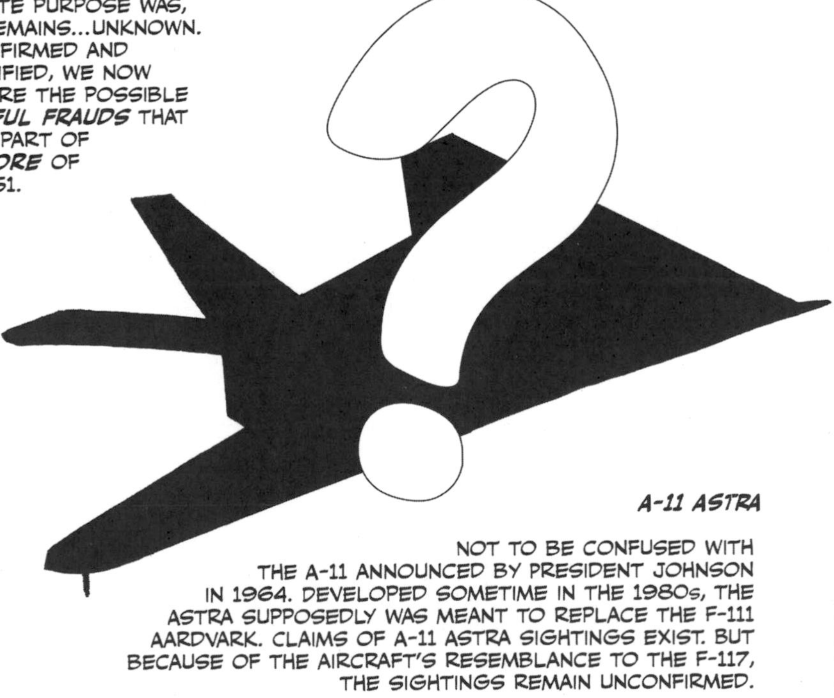

A-11 ASTRA

NOT TO BE CONFUSED WITH THE A-11 ANNOUNCED BY PRESIDENT JOHNSON IN 1964. DEVELOPED SOMETIME IN THE 1980s, THE ASTRA SUPPOSEDLY WAS MEANT TO REPLACE THE F-111 AARDVARK. CLAIMS OF A-11 ASTRA SIGHTINGS EXIST. BUT BECAUSE OF THE AIRCRAFT'S RESEMBLANCE TO THE F-117, THE SIGHTINGS REMAIN UNCONFIRMED.

F-121 SENTINEL

REPORTEDLY A SINGLE-SEAT MACH 3+ STEALTH RECONNAISSANCE AIRCRAFT FIRST FLOWN IN 1986. BELIEVED TO EXIST ON THE DRAWING BOARD. IT IS UNKNOWN IF ANY PROTOTYPES WERE BUILT.

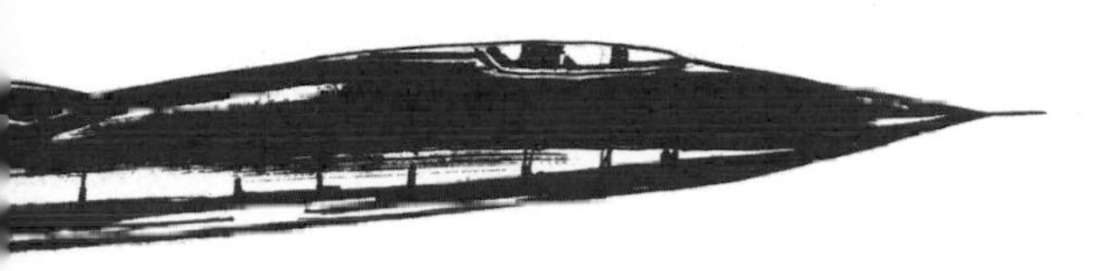

BRIGHT STAR

THIS MACH 2+ AIRCRAFT WAS APPARENTLY DESIGNED IN THE LATE 1980s BY SKUNK WORKS TO RESEARCH QUIET SUPERSONIC PLATFORM (QSP) CAPABILITY APPARENTLY PHOTOGRAPHED IN 1995 BY ANDREAS VON RETYI FROM TIKABOO PEAK, 26 MILES EAST OF AREA 51, THE CLOSEST LEGAL VANTAGE POINT FOR OBSERVANCE. IT IS DISTINGUISHED BY THE *"DOUGHNUTS-ON-A-ROPE"* CONTRAIL. IDENTITY OF THE AIRCRAFT IN RETYI'S PHOTOGRAPHS HAS NOT BEEN VERIFIED.

HYPERSONIC GLIDE VEHICLE (HGV)

A RECOVERABLE, UNMANNED, ROCKET-POWERED HYPERSONIC AIRCRAFT CAPABLE OF MACH 18. ALSO BELIEVED CAPABLE OF CARRYING NUCLEAR WARHEADS. ACCORDING TO UNCONFIRMED REPORTS, DESIGNS DATING FROM 1979 EXIST. ACTUAL TEST FLIGHTS HAVE NOT BEEN PUBLICIZED.

SNOW BIRD/BRILLIANT BUZZARD

A MACH 3.5+ RECONNAISSANCE AIRCRAFT SUPPOSEDLY DESIGNED TO SUCCEED THE SR-71. ITS NAME APPARENTLY IS IN REFERENCE TO ITS WHITE THERMAL PROTECTION TILES. INITIAL TEST FLIGHTS APPARENTLY WERE CONDUCTED IN 1999. WHILE SEVERAL SUPERSONIC DELTA-WINGED SUCCESSORS TO THE SR-71 ARE KNOWN TO EXIST IN VARYING STAGES, NONE HAVE YET TO BE ACKNOWLEDGED.

L301/COPPER COAST

REPORTEDLY A NASA-LINKED PROJECT BEGUN IN THE MID-1970s, THE L301/COPPER COAST WAS A HYPERSONIC AIRCRAFT CAPABLE OF MACH 6.65. THIS DELTA-WINGED AIRCRAFT WAS CANCELED AND THEN REPORTEDLY REVIVED AS A MACH 4 AIRCRAFT. THOUGH THE ORIGINAL DESIGN IS KNOWN TO EXIST, THERE IS NO AVAILABLE INFORMATION CONFIRMING A POSSIBLE REVIVAL.

AS WITH SO MUCH ELSE ABOUT AREA 51, UNTIL THE MILITARY AND INTELLIGENCE COMMUNITIES DECLASSIFY AND CONFIRM THESE--AND OTHER PROJECTS, THE TRUTH SURROUNDING THEIR EXISTENCE CAN ONLY BE THE SUBJECT OF SPECULATION.

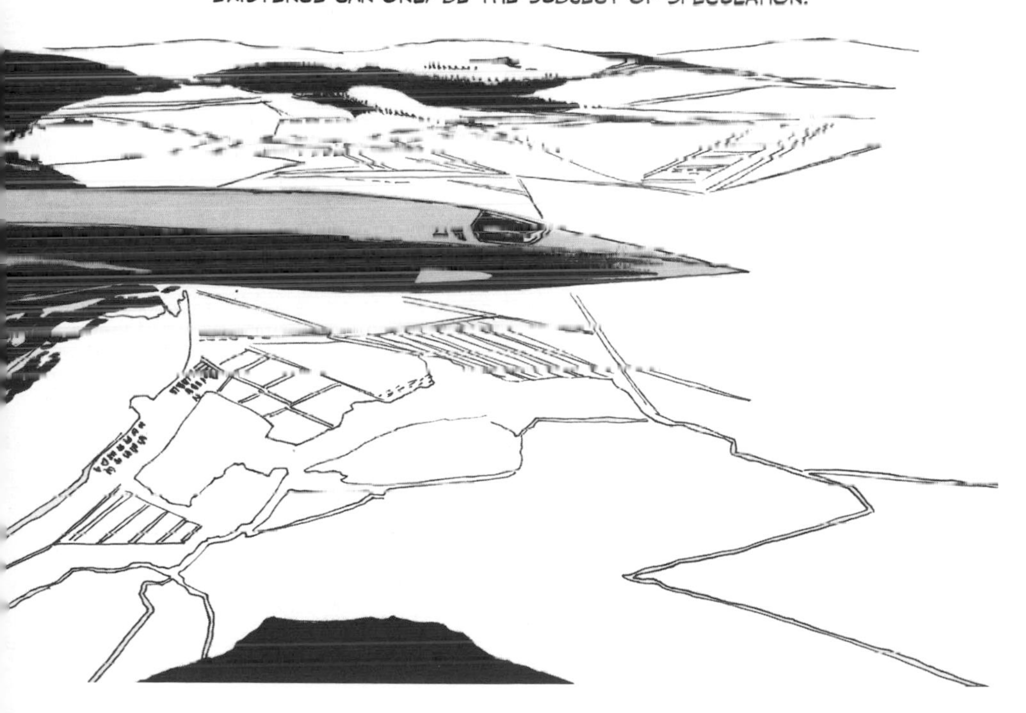

CHAPTER NINE
THE OUTING OF AREA 51
IN NOVEMBER 1989, IN A SERIES OF MEDIA INTERVIEWS, *ROBERT SCOTT LAZAR*, AN ENGINEER WHO HAD WORKED AT AREA 51, STUNNED AUDIENCES WITH HIS REVELATION OF THE SITE.
OKAY, WHAT'S GOING ON UP AT GROOM LAKE? CAN YOU TELL US THAT?
KUEG
THERE ARE ACTUALLY A LOT OF PROJECTS ARE GOING ON AT GROOM LAKE, AREA 51. ONE OF THEM IS AURORA ...
WHAT WAS YOUR FIRST REACTION THE FIRST TIME YOU KNEW FOR A FACT THAT WE HAD FLYING SAUCERS IN OUR POSSESSION...?
...THE FIRST TIME I SAW IT AND I WALKED IN AND ACTUALLY SAW THE DISK...IT WAS A LITTLE WHILE BEFORE I HAD ASCERTAINED THAT IT WAS AN EXTRATERRESTRIAL CRAFT.
LAZAR'S STORY WAS PICKED UP BY A JAPANESE TV SHOW AND HE TOLD A CREW WHERE TO GO ON TIKABOO MOUNTAIN TO FILM ACTIVITY ON AREA 51.
THE FILM AND THE STORY OF AREA 51 WERE SEEN BY 30 MILLION VIEWERS IN JAPAN. FROM THERE, THE STORY SNOWBALLED.
LAZAR PROVED TO HAVE A CONTROVERSIAL PAST, WHICH CAUSED HIM TO BE DISCREDITED.
THEY WERE NOT DISAPPOINTED. SHORTLY AFTER SUNSET, THEY SAW THEIR FLYING SAUCER.
THOUGH THE GOVERNMENT CONSISTENTLY DENIED ITS EXISTENCE, THE PUBLIC WAS NOW, AT LEAST UNOFFICIALLY, AWARE OF AREA 51.

IN ADDITION TO BEING AN ACTRESS, ***SHIRLEY MacLAINE*** WAS ALSO KNOWN FOR HER INTEREST IN REINCARNATION, UFOs, AND THE PARANORMAL.

NOW, WHEN YOU FIRST BECOME PRESIDENT, ONE OF THE FIRST QUESTIONS PEOPLE ASK YOU IS...
"WHAT'S REALLY GOING ON IN AREA 51?"
THE AUDIENCE REACTED WITH LAUGHTER.
WHEN I WANTED TO KNOW, I CALLED SHIRLEY MACLAINE.
I THINK I JUST BECAME THE FIRST PRESIDENT TO EVER PUBLICLY MENTION AREA 51.

HOW'S THAT, SHIRLEY?

CENTRAL INTELLIGENCE AGENCY
UNITED STATES OF AMERICA
CIA ADMITS "AREA 51" EXISTENCE
PRESIDENT OBAMA WAS CORRECT. BUT BY THEN, SUCH A STATEMENT HAD BECOME MOOT.
ON AUGUST 16, 2013, THE CIA RELEASED DOCUMENTS THAT OFFICIALLY IDENTIFIED AREA 51.

LEGAL PRESSURE HAD BEEN BUILDING FOR YEARS THAT ALSO THREATENED TO FORCE THE GOVERNMENT TO ACKNOWLEDGE AREA 51'S EXISTENCE.
SINCE 1970, THE OCCUPATIONAL SAFETY AND HEALTH ADMINISTRATION, AS WELL AS A SERIES OF LAWS, REGULATED THE EXPOSURE OF WORKERS TO TOXIC AND HAZARDOUS WASTES AND THEIR DISPOSAL.

SPECIAL INCINERATORS WERE DESIGNED TO SAFELY DISPOSE OF SUCH WASTE.

BUT NO SUCH FACILITY EXISTED AT AREA 51. DURING THE 1970s AND 1980s, THE TOXIC MATERIALS WERE DUMPED INTO DITCHES ...
... AND BURNED IN THE OPEN.
IN THE LATE 1980s AND EARLY 1990s, WORKERS EXPOSED TO THE FUMES BEGAN EXPERIENCING HEALTH PROBLEMS.
TESTS REVEALED HIGH LEVELS OF DIOXINS AND OTHER CARCINOGENS.
LAWSUITS WERE FILED BY WORKERS AND THEIR FAMILIES.
A FEDERAL JUDGE DISMISSED THE LAWSUITS, CITING NATIONAL SECURITY PRIVILEGE. THE CASES WERE APPEALED ALL THE WAY TO THE SUPREME COURT, WHICH REFUSED TO REVIEW THEM.

CHAPTER TEN

THE FUTURE OF AREA 51

IT'S TOO SOON TO TELL WHAT SORT OF IMPACT--IF ANY--GOVERNMENT ACKNOWLEDGMENT OF AREA 51 WILL HAVE.

AFTER ALL, THE RUSSIANS, WHO WERE THE PRIMARY TARGET OF AREA 51'S WEAPONS TESTING, HAVE BEEN MONITORING ITS ACTIVITIES SINCE THE 1960s.

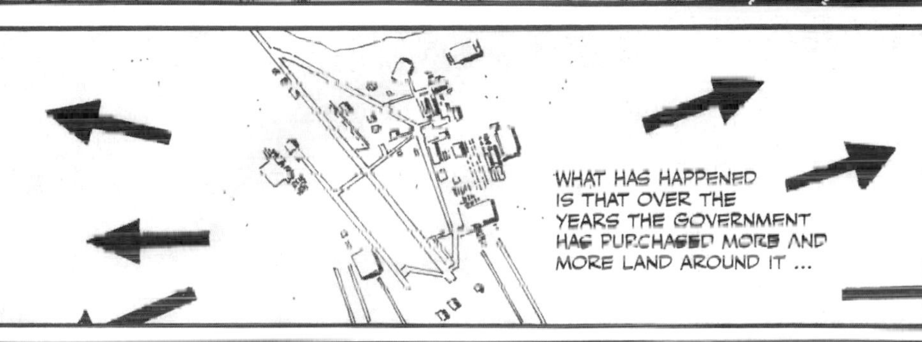

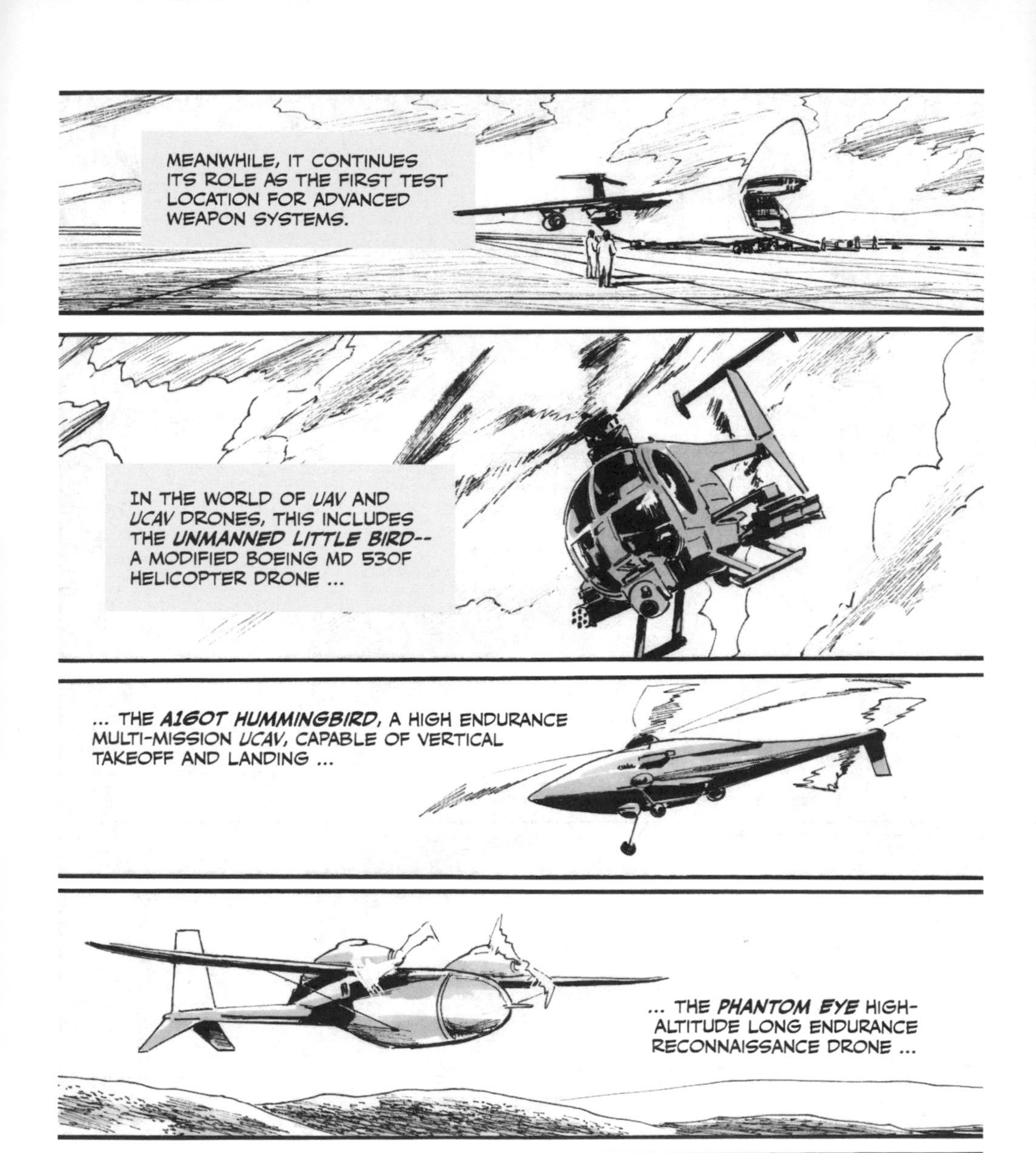
MEANWHILE, IT CONTINUES ITS ROLE AS THE FIRST TEST LOCATION FOR ADVANCED WEAPON SYSTEMS.
IN THE WORLD OF UAV AND UCAV DRONES, THIS INCLUDES THE UNMANNED LITTLE BIRD-- A MODIFIED BOEING MD 530F HELICOPTER DRONE ...
... THE A160T HUMMINGBIRD, A HIGH ENDURANCE MULTI-MISSION UCAV, CAPABLE OF VERTICAL TAKEOFF AND LANDING ...
... THE PHANTOM EYE HIGH-ALTITUDE LONG ENDURANCE RECONNAISSANCE DRONE ...

... AND THE BOEING/INSITU GROUP SCANEAGLE INTELLIGENCE UAV.

AND, IN THE REALM OF HYPERSONIC AIRCRAFT, THERE ARE:
THE DARPA FALCON PROJECT (FORCE APPLICATION AND LAUNCH FROM CONTINENTAL UNITED STATES), A BOMBER CAPABLE OF FLYING AT A SPEED OF MACH 20 ...
DARPA
SR-72
... AND SKUNK WORKS' SUCCESSOR TO THE SR-71, THE LOCKHEED-MARTIN SR-72, CAPABLE OF MACH 6 SPEED.

AS FOR WHAT ELSE IS GOING ON IN AREA 51, THAT INFORMATION IS *DENIED*.
TOP SECRET SPECIAL ACCESS PROGRAMS
YOU ARE *NOT CLEARED* FOR ACCESS TO MATERIAL CLASSIFIED AS *NEED-TO-KNOW*.